Elementary Photochemistry with Solutions

Elementary Photochemistry with Solutions

Manoj Kulkarni

CYBER TECH PUBLICATIONS
4264/3, Ansari Road, Daryaganj, New Delhi-110002 (India)
Ph.: 011-23244078, 011-43559448 Fax: 011-23280028
E-Mail: cyberpublicationsdelhi@yahoo.com
Website: www.cybertechpublications.com

ELEMENTARY PHOTOCHEMISTRY WITH SOLUTIONS

Manoj Kulkarni

First Edition 2010

Asian Offset Printers

Published by :

G.S. Rawat for **Cyber Tech Publications**
4264/3, Ansari Road, Daryaganj, New Delhi-110002 (India)
Ph.: 011-23244078, 011-43559448 Fax: 011-23280028
E-Mail: cyberpublicationsdelhi@yahoo.com
Website: www.cybertechpublications.com

PREFACE

This book "Elementary Photochemistry with Solutions" is provides all necessary information. The present edition has many new features. This approach is adopted for some main reasons. Several sections in all these chapters have been rewritten and enlarged to make the text clear. To familiarize the student about expected type of questions in the examinations, important summary and review questions have been provided at the end of each chapter. The subject matter has been presented in a simple and systematic manner. So, the students written some querries and mistakes in this book.

Sufficient problems has also been selected from various universities examination papers. The proofs of various principles and examples has been given with minute details. The author shall be gratefully to the readers who point out errors and omissions which inspite of all care might have been there. While this was gratifying in one sense, at the same time there were many imitating features like typographical errors.

— Author

Contents

1

Photochemical Reactions of Coordination Compounds

INTRODUCTION

The preceding discussion has emphasized that an electronic excited state of a coordination compound is an energy–rich species whose structure, electronic distribution and reactivity can differ substantially from the corresponding properties of the ground state. In this section the reactions of the various types of excited states are considered in detail. Most attention will be given to six–coordinate complexes of metals having a d^3 or low-spin d^6 electronic configuration, since the majority of mechanistic studies reported to date are concerned with these systems. Complexes of other d transition elements, the lanthanides and the actinides will also be treated, but the coverage here will necessarily be brief.

LIGAND FIELD EXCITED STATES

To a first approximation, the ligand field excited states of coordination compounds arise from electronic transitions between valence d-orbitals localized on the metal. While such transitions result in an angular rearrangement of metal electron density, they do not alter the net charge distribution between the metal and its ligands.

Consequently, ligand field excited states are not expected to undergo intramolecular oxidation-reduction processes such as homolytic fission of a metal-ligand bond. Angular changes in electron density can, however, lead to enhanced reactivity toward ligand substitution and isomerization processes. This theme will be developed in detail for a number of well-studied transition metal systems.

CHROMIUM(III) COMPLEXES

The low-lying electronic states of six-coordinate Cr^{III} complexes in octahedral (O_h) symmetry. The $^4A_{2g}$ ground state corresponds, in the strong-field limit, to the t_{2g}^3 electronic configuration whereas the $^4T_{2g}$ and $^4T_{1g}$ (F) ligand field excited states correlate with $t_{2g}^2e_g^1$. Consequently, transitions to these latter states occur with the promotion of an electron to the strongly σ–antibonding e_g orbitals. This redistribution of electron density weakens metal-ligand bonding and can cause considerable distortion of the complex from its original octahedral framework. Moreover, the vacated t_{2g} orbital is available for bonding to an incoming ligand.

These characteristics of the $^4T_{2g}$ and $^4T_{1g}$ (F) excited states make them particularly susceptible to ligand substitution reactions. In contrast, the lowest doublet excited state, ^2E_g, arises from a spin-pairing of two electrons within the t_{2g} orbitals. Since these orbitals are nonbonding (except with respect to π–donor or acceptor ligands), this change in electronic distribution should not cause major alterations in metal-ligand bond strength or overall geometry. A recent theoretical study shows, however, that the activation barrier to associative attack by an incoming nucleophile should be lower in ^2E_g than in the ground state. Thus, the lowest doublet state in Cr^{III} complexes must also be considered as potentially reactive toward ligand substitution.

In accord with these expectations, the vast majority of Cr^{III} complexes undergo ligand substitution when irradiated in their ligand field absorption bands. Generally, the leaving ligand is replaced by solvent, although anation may occur at high anion concentration. Complexes having six equivalent metal-ligand bonds, such as $[Cr(NH_3)_6]^{3+}$, undergo initial photosubstitution of a single ligand. The analogues process for $[Cr(en)_3]^{3+}$ (equation (2) results in the detachment and protonation of one end of an ethylenediamine ligand (denoted as enH). Thermal cleavage of the second Cr–N bond then yields a mixture of *cis*–and *trans*–$[Cr(en)_2(OH_2)_2]^{3+}$.

$$[Cr(NH_3)_6]^{3+} \xrightarrow[H_3O+]{h\nu} [Cr(NH_3)_5(OH)_2]^{3+} + NH_4^+ \quad ...(1)$$

$$[Cr(en)_3]^{3+} \xrightarrow{h\nu} [Cr(en)_2(enH)(OH_2)]^{4+} \quad ...(2)$$

A pulsed-laser photolysis study of $[Cr(en)_3]^{3+}$ illustrates quite dramatically the enhancement in reactivity that can result upon populating a ligand field excited state. Thus, a significant fraction of the primary photoproduct, $[Cr(en)_2(enH)(OH_2)]^{4+}$, is formed within the 20 ns duration of the laser pulse and in thought to arise from reaction of the lowest excited quartet state, $Q_1°$. This observation establishes that the pseudo-first-order rate constant for this excited state process is > $10^8 s^{-1}$. By comparison, the rate constant for the corresponding ground-state thermal reaction is 8×10^{-7} s^{-1}at 25°C. Thus, the promotion of a single electron from the t_{2g} to the e_g orbitals leads, in this case, to a 10^{14} acceleration in the rate of ligand substitution.

Mixed-ligand Cr^{III} complexes have a particularly rich substitutional photochemistry in that two (or more) reaction modes are normally observed. The dominant photochemical reaction for $[CrX(NH_3)_5]^{2+}$ complexes in aqueous solution is NH_3 aquation, with X aquation to a lesser extent. In contrast, the latter pathway is the favoured thermal reaction of these compounds. Such behaviour again illustrates that the reactivity of ligand field excited states can differ sharply from that of the ground state. One of the most controversial questions encountered in studies of Cr^{III} photochemistry concerns the identity of the reactive excited state (s). As illustrated earlier, uncertainty can arise because the Franck-Condon state initially populated during irradiation in the quartet absorption bands generally undergoes rapid equilibration to the lowest the xi quartet state, $Q_1°$, and/or intersystem crossing to the lowest doublet, $D_1°$. Reaction from either or both of the latter states must therefore be considered.

In a very important study of the quartet-doublet reactivity question, it was found that ligand-field irradiation of $[Cr(CN)_6]^{3-}$ in degassed dimethylformamide results in both phosphorescence from the doublet and substitution of cyanide. In air-saturated solution, however, phosphorescence is quenched completely while the photoreaction is unaffected. The two processes appear to be uncoupled and thus, originate from different excited states. The most straightforward conclusion is that the lowest quarter excited state in $[Cr(CN)_6]^{3-}$ is the sole precursor to photoreaction.

Studies of the photochemistry of $[Cr(bipy)_3]^{3+}$, $[Cr(phen)_3]^{3+}$ and various substituted analogues have revealed behaviour quite different from that just described. In the prototypal case of $[Cr(bipy)_3]^{3+}$,ligand-field irradiation of the complex in neutral or basic aqueous solution results in phosphorescence and loss of one bipyridine ligand. Intersystem crossing from the lowest quartet state to the luminescent doublet occurs with near unit efficiency, while back-intersystem crossing is too endothermic to be of importance. Addition of 0.1 M iodide ion quenches

> 99.9% of the phosphorescence and ~98% of the photochemistry. This parallel behaviour of the two processes strongly suggests that the quenchable portion of the reaction originates from the doublet.

The issue of quartet *versus* doublet reactivity becomes considerably more complicated for most other Cr^{III} complexes such as $[Cr(en)_3]^{3+}$ *trans*–$[Cr(NCS)_4(NH_3)_2]^-$ and *trans*–$[Cr(NCS)_2(en)_2]^+$. Both phosphorescence and photochemistry can be quenched in these systems, with the former process being affected to a greater degree. While the unquenchable portion of the photochemistry most likely involves prompt reaction from the lowest excited quartet state, the quenchable component could result from either (*i*) delayed reaction from the quarter following back-intersystem crossing from the doublet, or (*ii*) reaction from the doublet state itself. Despite a rather extensive series of investigations, the question as to which pathway (i or ii) dominates is still contentious.

At present, then the situation with regard to the identity of the reactive excited states (s) in Cr^{III} substitutional photochemistry can be summarized as follows : The lowest quartet excited state is the sole reactive state in $[Cr(CN)_6]^{3-}$ and is at least partly responsible for the photochemistry of several other CrIII complexes. The lowest doublet state appears to be the predominant reactive state in $[Cr(bipy)_3]^{3+}$ and similar tris(polypyridyl) chromium(III) complexes and may also be reactive in a number of other complexes, Quite clearly, this aspect of Cr^{III} photochemistry requires additional study.

The photochemical and thermal substitution reactions of Cr^{III} complexes can follow quite different stereochemical courses. The behaviour described by equation (4) is typical : thermal aquation of *trans*–$[CrCl_2(en)_2]^+$ proceeds with retention of configuration, whereas photoaquation is accompanied by a highly stereospecific (≥ 99% *cis* product) rearrangement. With very few exceptions, it has been found that stereochemical mobility is a general feature of Cr^{III} photosubstitution reactions.

An interesting correlation between the ease of stereomobility during photolysis and the efficiency of photosubstitution has been observed for complexes of the type *trans*$[CrCl_2N_4]^+$ (Table 1 N_4 is $4NH_3$, 2en, 2,3,3,–tet or cyclam). While the overall ligand field strength is approximately constant within the series, the quantum yield for chloride aquation drops by 10^3 as the ammine is varied from NH_3 to cyclam. This decrease parallels the increasing chelation about the metal atom. The cyclam complex, in which the metal is completely surrounded by the macrocyclic tetradentate ligand, is remarkably photoinert. Significantly, it is also the only member of the series which cannot undergo stereochemical rearrangement and thus, yields the *trans*–chloroaquo complex as the major photoproduct. Though the data base is limited, these results suggest that stereomobility is a necessary condition for efficient substitutional photochemistry in Cr^{III} complexes.

An angular overlap analysis of steremobility in Cr^{III} photosubstitution reactions has been presented by Vanquickenborne and Ceulemans. In this treatment, the overall reaction is formally dissected into three steps:

(1) dissociative loss of a ligand from the lowest quartet excited state of the complex,

(2) symmetry-controlled rearrangement of the resulting electronically excited square pyramid to a trigonal bipyramid.

(3) symmetry-controlled attack of the incoming ligand on the trigonal bipyramid to yield final product. Extensive comparisons between the stereochemical predictions of this theory and experiment have been presented and, in general, the agreement is satisfactory. Nevertheless, the assumption of dissociative ligand loss (*i.e.*, step 1 is not entirely compatible with other evidence. For example, the quantum yields for anation of $[Cr(DMSO)_6]^{3+}$ by azide are larger than those fro anation by thiocyanate at all concentrations, including those that approach limits imposed by ion pairing. This dependence on entering ligand, coupled with the observation that photoanation proceeds more efficiently than photoinduced solvent exchange, is indicative of an associative reaction pathway. In another study, the pressure dependence of photoaquation was determined for $[CrX(NH_3)_5]^{2+}$ (X is Cl, B or NCS) and $[Cr(NH_3)_5]^{3+}$. The negative volumes of activation found for both NH_3 and X^-loss were interpreted in terms of an I_a (associative interchange) mechanism. The extent of dissociative *versus* associative behaviour in Cr^{III} photosubstitution reactions is an interesting and important question that will undoubtedly receive more attention in the future.

Thus, far the discussion has focused on the reactivity of the lowest excited state of a given spin multiplicity (*e.g.*, $^4T_{2g}$ or ^2E_g in O_h symmetry) since, as noted earlier, higher lying states generally undergo rapid deactivation by intersystem crossing and/or internal conversion. In some complexes, however, it appears that reaction from upper ligand field thexi states may be competitive with these latter processes. The behaviour of *trans*–$[CrF(NCS)(en)_2]^+$ is illustrative in this regard.

The complex undergoes photosubstitution of both thiocyanate and ethylenediamine with a quantum yield ratio, ϕNCS-/ϕen, that varies with the wavelength of irradiation. This wavelength dependence requires the existence of two reactive excited states, and it was proposed that the $^4E(a)$ and $^4B_2$ states in this nominally C_{4v} complex are involved. The $^4E(a)$ state gives rise to both en and NCS–labillization whereas the higher energy $^4B_2$ state favours en loss. Extension of the Vanquickenborne-Ceulemans photochemical theory to include the reactivity of higher lying quartet excited states leads to predictions of reactivity which are in accord with the observed behaviour.

There are now several reports of a small wavelength dependence of photochemical and emission quantum yields on irradiation into the red edge (long-wavelength portion) of the lowest quarter ligand field absorption band. Such behaviour has been attributed to prompt intersystem crossing and/or reaction occurring from vibronic levels at or near the Franck-Condon state initially populated during light absorption.

These processes must be extremely rapid (*i.e.,* picosecond time scale) to be able to compete successfully with vibrational relaxation to thexi state. A theoretical treatment of prompt photoreactions has recently been presented. Finally, it should be noted that the relatively long-lived doublet let excited state in Cr^{III} complexes can undergo bimolecular processes. Energy transfer is a well-documented pathway for this state and, more recently, electron transfer reactions have also been observed. In the example given the lowest doublet of $[Cr(bipy)_3]^{3+}$ undergoes reductive quenching by $[RuI(bipy)_3]^{2+}$. While this redox process is energetically unfavourable when both reactants are in their ground electronic states, the higher energy of the doublet state provides the necessary driving force to accomplish the reaction photochemically. Additional examples of bimolecular excited state electron transfer processes are presented in Section 2B.

$$(^2E)Cr(bipy)_3]^{3+} + [Ru(bipy)_3]^{2+} \longrightarrow [Cr(bipy)_3]^{2+} + [Ru(bipy)_3]^{3+}$$

COBALT(III), RHODIUM(III) AND IRIDIUM(III) COMPLEXES

The energy level diagram for the six-coordinate complexes of CoIII, RhIII and IrIII in Oh symmetry. The 1A1g ground state derives from the

t_{2g}^{6} electronic configuration, while the low-lying ligand field excited states 3T1g, ${}^3T_{2g}$, ${}^1T_{1g}$, and ${}^1T_{2g}$, correlate with the $t_{2g}^{5}e_g^{1}$ configuration. There is, in addition, a${}^5T_{2g}$ state which arises from the $t_{2g}^{4}e_g$ configuration and whose energy is very sensitive to the ligand field strength. It is generally agreed that the quintet state lies above the ${}^3T_{1g}$ state in Rh^{III} and Ir^{III} ammine complexes as well as the strong-field complexes of Co^{III} such as $[Ci(CN)_6]^{3-}$, whereas the relative positions of these states may be reversed for complexes of lower field strength such as $[Co(NH_3)_6]^{3+}$. This change in the nature of the lowest excited state may account for the different emission properties of the two types of complexes. Thus, Rh^{III} and Ir^{III} ammine and Co^{III} cyanide complexes typically exhibit a broad phosphorescence assigned as${}^3T_{1g} \longrightarrow {}^1A_{1g}$ while CO^{III} ammines are nonemissive even at low temperature. Table 1 summarizes the photochemistry of several Co^{III}, RhIII and Ir^{III} complexes upon ligand field excitation. Ligand substitution is the sole reaction mode and, more than one photoproduct may result for mixed-ligand complexes such as $[RhX(NH_3)_5]^{2+}$ (X is Cl, Br or I).

Table 1 : Photosubstitution Quantum yields for some d^6 Complexes.

Complex	λ_{irr}	*Assignment*	*ϕNH_3(or amine)*	
$[Co(NH_3)_6]^{3+}$	365	${}^1A_{1g}\rightarrow{}^1T_{2g}$	5.4×10^{-3}	-
	460	${}^1A_{1g}\rightarrow{}^1T_{1g}$	5.2×10^{-4}	-
$[Co(CN)_6]^{3-}$	365	${}^1A_{1g}\rightarrow{}^1T_{1g}$	-	0.31
	405	${}^1A_{1g}\rightarrow{}^3T_{1g}$	-	0.29
$[Rh(NH_3)_6]^{3+}$	254	${}^1A_{1g}\rightarrow{}^1T_{2g}$	0.07	–
	313	${}^1A_{1g}\rightarrow{}^1T_{1g}$	0.075	–
$[Ir(NH_3)_6]^{3+}$	254	${}^1A_{1g}\rightarrow{}^1T_{1g}$	0.083	–
	313	${}^1A_{1g}\rightarrow{}^3T_{1g}$	0.090	–
$[RhCl(NH3)5]^{2+}$	380	${}^1A_1\rightarrow{}^1T_1$	$<10^{-3}$	0.14
$[RhBr(NH_3)_5]^{2+}$	420	${}^1A_1\rightarrow{}^1E$	0.17	0.019
$[RhI(NH_3)_5]^{2+}$	420	${}^1A_1\rightarrow{}^1E$	0.87	0.01
trans–$[RhCl_2(NH_3)_4]^{+}$	407	a	$<2 \times 10^{-3}$	0.13
trans–$[RhCl_2(en)_2]^{+}$	407	a	$<3 \times 10^{-3}$	0.057
trans–$[RhCl_2(cyclam)]^{+}$	407	a	$<10^{-3}$	0.011

a Irradiation into first spin-allowed ligand field absorption band.

The data reveal an interesting dichotomy between the photoreactivities of Rh^{III} and Ir^{III} ammines and Co^{III} cyanides on the one hand and Co^{III} ammines on the other. The former complexes undergo reasonably efficient substitution and, in most cases, the quantum yields are independent of the irradiation wavelength (λ_{irr}).

Such behaviour indicates that photo-excitation is followed by efficient internal-coversion/intersystem crossing to a common excited state, logically assigned as the lowest ligand fields triplet, and that this state and those in thermal equilibrium with it are mainly responsible for the substitutional chemistry. This conclusion is reinformed by the finding that triplet-state sensitizers drive the reactions with limiting quantum yields that closely match those obtained upon direct photolysis. In sharp contrast, Co^{III} ammines undergo photosubstitution with extremely low and wavelength-dependent quantum yields.

Since ϕ values increase with higher-energy excitation, it can be concluded that much of the observed substitution chemistry arises from higher-lying ligand field excited states. The small photosensitivity of Co^{III} ammines, particularly the lowest-energy excited state, has been attributed to the weak ligand field in these complexes. As noted earlier, at sufficiently low field strength the energy of the quintet state, $^5T_{2g}$, falls below that of $^3T_{1g}$. In such cases configurational mixing between $^5T_{2g}$ and the ground state enhances radiationless decay and thus, chemical processes (as well as luminescence) will be strongly diminished.

The photosubstitution reactions of Co^{III} and Rh^{III} complexes can occur with a high degree of stereochemical change; some representative examples are presented in equations 17–19. Stereomobility during photolysis is not universal, however, as evidenced by the behaviour within the series of *trans*–$[RhCl_2N_4]^+$ complexes N_4 is $4NH_3$, 2en or cyclam). Photosubstitution of chloride occurs with complete retention of configuration (equation 5) and, significantly, the quantum yields for this process show much less sensitivity to increasing chelation about the metal that was found for the corresponding Cr^{III} complexes. Results such as these demonstrate that stereomobility is not a requirement for efficient photosubstitution in d^6 systems.

$$\textit{cis}\text{–or } \textit{trans}[CoCl_2(en)_2]^+ \xrightarrow[H_2O]{hv} [CoCl(en)_2(OH_2)]^{2+} + Cl^- \quad ...(5)$$

30% cis + 70% trans

$$cis\text{–}[RhCl_2(NH_3)_4]^+ \xrightarrow[H_2O]{hv} trans\text{–}[RhCl(NH_3)_4(OH_2)]^{2+} + Cl^- \quad ...(6)$$

$$\text{cis–or trans}[RhCl(OH)(NH_3)_4]^+ \xrightarrow[OH^-]{h\nu} \text{cis–}[Rh(OH_2)(NH_3)_4]^+ + Cl^- \quad ...(7)$$

$$\text{trans–}[RhCl_2N_4]^+ \xrightarrow[H_2O]{h\nu} \text{trans]}[RhClN_4(Oh)_2]^{2+} + Cl^- \quad ...(8)$$

The photochemical model developed by Vanquickenborne and Ceulemans provides a convenient theoretical framework for discussing the photosubstitutional behaviour of d^6 complexes. In this treatment it is assumed that reaction occurs from the lowest triplet state of the complex and that any stereochemical change takes place subsequent to ligand loss. These features the case of *cis*– and *trans*$[RhCl_2(NH_3)_4]^+$.

For both complexes, angular overlap considerations lead to the conclusion that relative bond strengths lie in the order Rh–N >Rh–Cl. Loss of $Cl^-$from the triplet state, ^{3}*cis,* of the *cis* isomer produces and electronically excited square-pyramidal intermediate, ^3SP_a, in which the remaining Cl^- lies in the basal plane. Similarly, loss of Cl^- from the triple state of the *trans* isomer generates ^3SP_b, which contains a Cl^- ligand in the apical position. The square pyramidal intermediates can (*i*) interconvert along are reaction coordinate that passes through a trigonal bipyramidal transition state of triplet spin multiplicity, ^{3}TBP, or (*ii*) relax to their respective ground states and add a molecule of solvent. The indentity of the final product therefore depends upon the competition between these two pathways. Calculation of relative energies indicates that the ^{3}SP intermediate of lowest energy will be the one having the weaker σ–donor ligand in the apical site and that the barrier for interconversion, ΔE, becomes larger with increasing ligand field strength.

Thus, in the example ^3SP_a is predicted to be less stable than ^3SP_b. Isomerization of the former to the latter, followed by deactivation to ^1SP_b and rapid trapping by solvent, would give rise to the observed *trans* product in equation (7). Apparently, 3SPb faces too high a barrier (ΔE_r) for conversion to ^3SP_a and, as a result, aquation of the *trans* complex in equation (8) occurs with retention of configuration. High barriers would also account for the observation that most Ir^{III} complexes undergo photosubstitution with stereoretention. The identical product mixtures formed by *cis* and *trans*-$[CoCl_2(en)_2]^+$ on the other hand, would require that the barriers for interconversion are sufficiently low (because of the smaller ligand field strengths of Co^{III} *versus* Rh^{III} complexes) to allow the same steady-state ratio of ^3SP_a and ^3SP_b to be reached by both isomers prior to deactivation and trapping by solvent. While ligand field considerations of the type discussed above usually lead to reliable predictions of photosubstitutional behaviour, it is important to note that they are not the only determinant of excited state reactivity. Environmental factors can also play an important role, especially in cases where ligand field effects do not strongly favour one reaction pathway over another. The behaviour of $[RhCl(NH_3)_5]^{2+}$ in different solvents illustrates this point, the predominant photoreaction changes from Cl^- loss in water or formamide to NH_3 loss in dimethyl sulfoxide, methanol or dimethylformamide. Measurements of the excited-state rate constants, K_{cr-} and K_{NHi}, for these two processes provide some useful insight about the intimate mechanism of ligand substitution.

Thus, K_{cr-} is found to be much more sensitive than is K_{NH3} to changes in solvent. Moreover, while neither rate constant correlates with the donor properties of the solvent, K_{cr-} qualitatively parallels the solvation energy of the chloride ion. These observations are best interpreted in terms of a dissociative mechanism in which a considerable degree of charge separation develops in the transition state for RH–Cl bond breaking and the overall energy for this process is influenced by the ability of the solvent to stabilize the developing charge. Other evidence, summarized elsewhere, supports the view that ligand field excited states of d^6 complexes generally undergo ligand substitutional *via* a dissociative process.

CHARGE TRANSFER EXCITED STATES

The various categories of CT transitions in coordination compounds were defined in section IB. Such transitions occur with a radical redistribution of electron density and thus, cause changes in the oxidation states of the species (metal, ligands, solvent) involved. This property of CT transitions leads to the expectation that the resulting excited states will be particularly susceptible to oxidation-reduction reaction. Ligand substitution pathways also may become important, especially in systems where charge transfer creates a substiutionally labile metal centre. Likewise, changes in the charge distribution about a coordinated ligand can enhance its reactivity toward processes such as protonation, electrophilic attack and isomerization.

In the acidoammine complexes of Cr^{III}, Co^{III} and Rh^{III}, CT states generally are higher in energy than other types of excited states. This situation for the case of $[Cr(NCS)NH_3)_5]^{2+}$, where the NCS $\rightarrow$ Cr CT state is seen to lie above intraligand and ligand field states. Since the CT state reached in absorption can, in principle, relax to any lower energy state, it is often difficult to assign the excited state responsible for a particular reaction. The occurrence of ligand substitution in $[Cr(NCS)NH_3)_5]^{2+}$ following irradiation of the CT band, for example, could result from reaction of the CT state and/or a lower-lying ligand field state.

Another factor that can complicate the interpretation of CT photochemistry is the highly reactive nature of the primary photoproducts. CT excited states normally react to produce radicals which can undergo secondary reactions with each other, the original complex or the solvent. Thus, the final products observed may not be a reliable indicator of the nature of the efficiency of the primary photochemical step. Fast-reaction techniques such as laser flash photolysis and pulse radiolysis have been extremely useful in helping to unravel the complex sequence of reactions that can occur in these systems.

LIGAND-TO-METAL CHANGE TRANSFER STATES

Most of the available information about the reactivity of CTTM excited states comes from studies of Co^{III} acidoammines. Typically, irradiation of a CTTM absorption band in these systems leads to efficient redox decomposition accompanied, in many cases, by ligand substitution or linkage isomerization. The behaviour of a representative complex, $[CoCl(NH_3)_5]^{2+}$, is summarized in equation (1). A common observation in these systems is that photoredox quantum yields increase at shorter excitation wavelengths and increasing solvent viscosity. The photochemistry of Co^{III} acidoammines upon CTTM excitation can be understood in terms of a mechanism in which homolytic fission of a metal-ligand bond is the primary reaction. This process, which produces a solvent-caged or primary radical pair, is shown in equation (2) for the case of $[CoX(NH_3)_5]^{2+}$ (X is Cl, Br, NO_2, NCS). The quantity Δ represents the amount of absorbed light energy in excess of that required for electron transfer, and it determines to a large extent the magnitudes of the radical fragments. When Δ is small, recombination of the cage partners is favourable equation 3(a), while for large Δ the radicals are more likely to diffuse apart to form a solvent-separated (S denotes solvent) or secondary radical pair . The secondary radical pair, in turn, can recombine to yield the original complex or diffuse apart with or without transfer of an electron back to X. The $[Co(NH_3)_5]^{2+}$ fragment produced in equation 4 is substitutionally labile and rapidly undergoes NH_3 loss to give aquated Co^{2+}.

According to this mechanism, photoredox quantum yields increase at shorter excitation wavelengths because of the increasing value of Δ and, hence, the greater likelihood of forming secondary radical pairs. Moreover, a drop in quantum yield with increasing solvent viscosity is expected, since the rates of the processes. Some modifications to the CTTM photochemistry of Cr^{III} acidoammine complexes. Thus the primary photochemical step in formation of a substitutionally labile reduced metal species and an oxidized ligand radical. In most systems, however, no permanent redox chemistry occurs owing to the facile reoxidation of the metal. The only net photoprocess observed in these cases is substitution of one or more ligands.

Bimolecular reactions of CTTM excited states are rare, presumably because of the rapid formation of the primary radical pair in equation (2). In principle, however, it should be possible to intercept the excited state by a reactive species present at sufficiently high concentration in solution. Some evidence for this type of process comes from a study of the photochemistry of $[Mo_2Cl_8]^{4-}$, a complex containing a quadruple metal-metal bond.

In strongly acid solution, population of an excited state thought to be CTTM in character results in production of $[Mo_2Cl_8H]^{3-}$. The proposal was made that the increased electron density at the metal centres in this excited state facilitates Mo-H bond formation. Further studies are needed to test the validity of this interesting suggestion.

B Metal-to-Ligand Charge Transfer States

Detailed studies of the reactivity of CTTL excited states are of rather recent vintage and have tended to focus on complexes of d^6 metals. This type of state formally contains an oxidized metal centre and a reduced ligand radical. In $[Ru(NH_3)_5py]2+$, for example, the low energy Ru $\longrightarrow$ py CT state can be formulated as $[(NH_3)_5RuIII\ (py)]^{2+}$. Several possible reactions can be envisioned for such a species. An intermolecular redox process leading to photooxidation of the metal is one potential pathway, but no net production of RuIII results upon direct CTTL excitation.

Another possibility is electrophilic attack by solvent on the reduced pyridine ligand. In acidic D_2O solution, electrophilic H/D exchange of pyridine protons does occur but with a rather small quantum yield (~ 10–3). The only significant photochemical reaction observed in this system is ligand substitution.

It appears unlikely, however, that this pathway originates from the Ru $\longrightarrow$ py CT excited state, since low-spin RuIII complexes (recall the RuIII (py) formulation) are known to be relatively substitution inert. Instead, it has been suggested that reaction originates form a ligand field excited state formed by interconversion from the initially populated CTTL state. Support for this assignment was obtained from a study of the photosubstitutional chemistry of a series of $[Ru(NH_3)_5(pyX)]^{2+}$ complexes, where pyX is a substituted pyridine or related aromatic heterocycle. Systematic variations in the electron withdrawing properties of pyx or, for a given ligand, the solvent polarity were used to 'tune' the energy of CTTL state relative to the lowest ligand field excited state. When the CTTL state lies above the reactive ligand field state, efficient intersystem crossing/internal conversion of the former to the latter results in a relatively high quantum yield for substitution. If the energies of the two states are reversed, however, deactivation leads to population of the unreactive CTTL excited state and the observed quantum yields are low. Similar behaviour has been reported for a series of $[Fe(CN)_5L]^{n-}$ complexes (n = 2, 3), where L is an aromatic nitrogen heterocycle.

While intramolecular photoredox reactions of CTTL excited states appear to be relatively unimportant, there is considerable evidence that these states can undergo bimolecular electron transfer with high quantum efficiency, examples of this type of process are given in equations (5) and (6) for the paradigm case, $[Ru(bipy)_3]^{2+}$, a complex possessing a long-lived Ru $\longrightarrow$ bipy CT excited state. The presence of an electron in a high energy π–antibonding orbital of a bipy ligand enhances the reducing ability of this state, while a vacancy (hole) in the t_{2g} metal orbitals increases its oxidizing ability. Thus, the excited complex is both a stronger reductant and a stronger oxidant than the corresponding ground state by the excitation energy of 2.1 eV. This behaviour is reflected in the reduction potential data shown in Figure 6.14 Several other transition metal complexes containing long-lived CTTL excited states undergo bimolecular electron transfer processes. The thermodynamic and kinetic aspects of these reactions and their potential applications in energy conversion schemes have been reviewed.

$$[Ru(bipy)_3]^{2+*} + Eu^{2+}_{(aq)} \longrightarrow [Ru(bipy)_3]^{+} + Eu^{3+}_{(aq)}$$

$$[Ru(bipy)_3]^{2+*} + Fe^{3+}_{(aq)} \longrightarrow [Ru(bipy)_3]^{3+} + Fe^{2+}_{(aq)}$$

There are also numerous examples of bimolecular energy transfer processes involving CTTL excited states. The Ru $\longrightarrow$ bipy CT state in $[Ru(bipy)_3]^{2+}$, for example, photosensitizes the reactions of a number of organic and inorganic substrates by this pathway.

Finally, it is conceivable that a CTTL excited state can be eject into the solution medium. In systems where this process has been observed, however, it has proved difficult to decide whether the reactive state is CTTL or CTTS in character. A discussion of photoelectron production is presented in the following section.

Charge-transfer-to-solvent states

CTTS transitions in coordination compounds result in a radial movement of electron density from the metal to the surrounding solution medium. The energies of these transitions generally are very sensitive to environmental parameters such as solvent polarity, temperature and the presence of salts.

This sensitivity has been used in a diagnostic sense to identify CTTS bands in the spectra of anionic cyanide complexes and 1, 2– dithiolene complexes of Ni, Pd and Pt. Hydrated cations such as $Cr^{2+}_{(aq)}$ and $Fe^{2+}_{(aq)}$ exhibit absorption bands that are sometimes referred to as CTTS in character. Since the solvent occupies the first coordination sphere of the metal, however, the distinction between CTTS and CTTL transitions in these systems becomes obscured.

A characteristic reaction of CTTS excited states is photoelectron production with concomitant oxidation of the metal centre. In the example given in equation (7) electrostatic repulsion of the primary photoproducts facilitates their separation and allows direct observation of the solvated in the presence of an electron scavenger such as N_2O or $CHCl_3$.

$$(CTTS)[Ru(CN)_6]^{4-*} \xrightarrow[H_2O]{} [Ru(CN)_6]^{3-} + e^-_{(aq)}$$

$$M^{n+}_{(aq)} + H^+ \xrightarrow{hv} M_{(aq)}^{(n+1)+} + \frac{1}{2} H_2$$

UV irradiation of low-Valent cations (*e.g.*, V^{2+}, Cr^{2+}, Fe^{2+}) in acidic aqueous solution leads to photooxidation of the metal and evolution of H_2. This reaction, described by equation (8), is thought to involve the formation of the oxidized metal and an electron in the primary photochemical act. Immediate scavenging of the photoelectrons by protons yields hydrogen atoms which then combine to produce H_2. Some interest has been shown in this type of systems as a component of a photochemical water-splitting cycle.

INTRALIGAND EXCITED STATES

Intraligand (IL) excited states of co-ordination compounds arise from electronic transitions between molecular orbitals primarily localized on a coordinated ligand. It is difficult, *a prior,* to predict the reactivity of this type of state. While it is logical to expect ligand-centered reactions, the influence of the metal on such processes can be substantial and result in net photochemistry which differs from that of the free ligand. A few examples should serve to illustrate the range of IL photoreactions reported to date.

The complex $[Co(TSC)(NH_3)_5]^{2+}$, where TSC^- denotes *trans*- stilbene carboxylate, exhibits an absorption band which is virtually identical to the spin-allowed π-π* band of free TCS^- and thus, is straightforwardly assigned as IL in character. The primary photoreaction upon irradiation into this IL band is redox decomposition to yield aquoted Co^{2+} and the radical resulting from oxidation of TSC. Photolysis of free TSC on the other hand, results in *trans–cis* isomerization.

One possible explanation for this disparate behaviour is that the initially populated singlet IL excited state of $[Co(TSC(NH_3)_5]^{2+}$ undergoes efficient intersystem crossing to a triplet CTTM excited state, from which redox decomposition then occurs. Alternatively, direct transfer of an electron from the locally excited TSC ligand to CoIII may be occurring. Regardless of the exact details of the mechanism, it is clear that the presence of the metal has a pronounced influence of the mechanism, it is clear that the presence of the metal has a pronounced influence on the reactivity of the IL state.

Metal perturbation of IL photochemistry appears to be less important in [Ru(bipy)2 (*trans*–4–stilbazole)$_2$]$^{2+}$ and [Ru(bipy)$_2$ (Cis–4–stilbazole)$_2$]$^{2+}$. Both complexes undergo wavelength-dependent isomerization of the coordinated stilbazole ligands as the only important photoreaction. This wavelength dependence has been attributed to the presence of two different types of reactive excited states, Ru $\longrightarrow$ stibazole CT state which favours formation of the *trans* isomer and a higher energy IL state localized on stilbazole which decays with nearly equal probability to the two isomers. The letter pathway is qualitatively similar to the isomerization process that obtains upon direct photolysis of the free ligand. Thus, co-ordination to Ru has not altered the characteristic photoreaction mode of the stilbazole molecule.

Photolysis of aqueous solution of $[RhN_3(NH_3)_5]^{2+}$ and $[IrN_3(NH_3)_5]^{2+}$ results in the evolution of N_2 and the formation of a coordinated nitrene (M–NH) intermediate. This cleavage of the N–N_2 bond originates primarily from an IL excited state localized on the azide ligand. The metal is thought to play an important chemical role in the excited state reaction by stabilizing the incipient nitrene through ad_π–p_π backbonding interaction.

SURVEY OF THE PHOTOREACTIONS OF D AND F TRANSITION METAL COMPLEXES

The preceding discussion of the relationships between excited state electronic structure and photochemical reactivity focused primarily upon coordination compounds containing d^3 or low spin d^6 transition metals. These relationships are generally applicable, however, to complexes of other d transition elements, the lanthanides and the actinides. A brief survey of the photochemical reactions of these latter systems is presented below.

Complexes of d Transition Elements

Several coordination compounds in which the central metal possesses a high oxidation state and a d^0 electronic configuration have been reported to undergo photoinduced homolytic cleavage of a metal-ligand bond, presumably the result of populating a CTTM excited state. UV or visible irradiation of $TiCl_4$, for example, causes homolysis of a Ti–Cl bond with the resulting formation of $TiCl_3$ and Cl_2. Vanadium complexes of general formula $VOQ_2(OR)$, where Q is the 8– quinolyloxo ligand and R is an alkyl group, undergo cleavage of the V–OR bond upon UV photolysis. This reaction occurs for a wide selection of R groups and thus, provides a convenient, general route to alkoxy radicals.

The eight-coordinate d^2 complexes $[Mo(CN)_8]^{4-}$ and $[WCN)_8]^{4-}$ undergo two wavelength dependent photoreactions. Photolysis in the short-wavelength (254nm) region leads to efficient photoelectron production from an excited state assigned as CTTS in character. Irradation into the longer-wavelength (365 nm) ligand field bands, on the other hand, favours ligand substitution. Tthe primary substituted photoproduct in basic media is the hydroxo complex which, in a series of thermal reactions, undergoes additional release of cyanide.

$$[M(CN)_8]^{4-} \xrightarrow[HO^-]{hv} [M(CN)_7(OH)]^{4-} + CN^-$$

In the dinuclear complex $[Re_2Cl_8]^{2-}$, the d^4 metal centres are linked by a quadruple bond. The lowest-lying singlet excited state in the complex arises from the δ–δ* transition of the Re_2 unit. This state has been found to undergo bimolecular electron transfer with a variety of substrates in acetonitrile solution, functioning either as a strong oxidant or a moderately good reductant.

Population of higher-energy excited states results in cleavage of the Re–Re bond and formation of monomeric $[ReCl_4(MeCN)_2]^-$. In the absence of evidence for direct photodissociation of the quadruple bond, it was proposed that this reaction proceeds by way of an intermediate in which a chloride ligand bridges the two metal atoms.

Since this type of structure leaves one of the Re atoms quite exposed to the environment, nucleophilic attack by solvent can assist the Re–Re rupture process.

Low spin d^5 complexes possess a CTTM excited state that arises from the transfer of a ligand electron to a metal orbital having predominantly nonbonding (t_{2g}) character. In $[RuCl(NH_3)_5]^{2+}$, for example, the Cl $\longrightarrow$ Ru CT state can be conceptualized as a Ru^{II} complex containing a bound ligand radical. Since the substitutional liability of Ru^{II} complexes is not especially great, this state might be expected to undergo internal back-electron transfer to regenerate the original complex faster than it decomposes to redox products. This expectation is borne out by the observation that observation that CTTM excitation of $[RuCl(NH_3)_5]^{2+}$ leads to relatively inefficient ($\phi = 0.02–0.04$) loss of NH_3 as the only net photoprocess. Similar considerations account for the lack of CTTM photo chemistry for $[IrCl_6]^{2-}$. Ligand field excitation of d_5 complexes generally results in ligand substitution and/or isomerization processes. As summarized, photolysis of *cis*-or *trans*–$[RuCl_2(en)_2]^+$ leads to an isomeric mixture of the chloroaquo products but, interestingly, the isomer distribution differs for the two starting complexes. This behaviour has been analysed in terms of the angular overlap model of ligand field photochemistry.

cis–$[RuCl(en)_2(OH)_2]^{2+}$

cis–$[RuCl_2(en)_2]^+$ $\xrightarrow[H_2O]{h\nu}$ 84% 66% $\xleftarrow[H_2O]{h\nu}$ *trans*–$[RuCl_2(en)_2]^+$

16% 34%

trans–$[RuCl(en)_2(OH_2)]^{2+}$

While four-coordinate d^8 complexes of Pt^{II} are square planar in the their ground electronic states, spectroscopic and theoretical studies suggest that the low-lying ligand field excited states are significantly distorted and may even approach tetrahedral geometry. Minimally, such distortion provides a ready pathway for isomerization, and can also result in labilization of specific metal-ligand bonds. Consider the example of Zeise's anion, $[PtCl_3C_2H_4]^-$, which in the ground state undergoes rapid aquation of the chloride *trans* to ethylene.

Ligand field excitation results is substitution of both *cis*-chloride and ethylene (the extent of *trans*–chloride photoaquation could not be determined). The photolability of ethylene was attributed to the distortion of ligand field excited states toward a tetrahedron, since in this geometry π–backbonding from Pt to C_2H_4 is weakened relative to that in the square planar ground state. Several Pt^{II} complexes undergo *cis–trans* isomerization when irradiated in their ligand field absorption bands, specific examples being $[PtCl_2(PEt_3)_2]$, $[Pt(gly)_2]$(gly is glycinate) and $[Pt(py)_2Cl_2]$.

The first two systems are thought to isomerize *via* an intramolecular twisting mechanism, whereas a combination of intramolecular twisting and intermolecular ligand substitution pathways is responsible for the rearrangement of the last complex.

In solution the d^8 complexes of Ni^{II} are characterized by several rapid equilibria between species having different coordination numbers and/or geometries. Upon irradiation, these equilibria are perturbed and can be studied by relaxation techniques. A number of photoinduced changes have been investigated in this manner including square planar/tetrahedral, four-coordinate/five-coordinate and four-coordinate/six coordinate interconversions.

The photochemistry of d^9 Cu^{II} and d^{10} Cu^I complexes is dominated by redox processes. Halide complexes of general formula $[CuCl_x]^{(2-x)-}$ or $[CuBr_x]^{(2-x)-}$ (x = 1–4), for example, undergo intramolecular electron transfer of the CTTM type to form a reduced metal species and a free halogen atom. CTTM photochemistry has also been observed for Cu^{II} complexes containing amines, carboxylates, β–diketonates or polypyridines.

Copper (I) halide complexes such as $[CuCl_3]^{2-}$ undergo photoelectron production, presumably the result of populating a CTTS excited state. In acidic aqueous media, scavenging of the electron by H^+ results in evolution of H_2.

Irradiation of $[Cu(dmp)_2]^+$ (dmpis 2, 9– dimethyl–1, 10-- phenanthroline) populates a low-lying CTTL excited state which can undergo bimolecular electron transfer to various Co^{III} complexes. Photoinduced intramolecular electron transfer from Cu^I to Co^{III} in a mixed-metal dimer has also been reported. Finally, [Cu(prophos) BH_4](phrophos is 1, 3–bis (diphenylphosphino) propane) has been shown to undergo bimolecular energy transfer to organic substrates in solution from and IL excited state localized on the prophos ligand.

Complexes of the Lanthanide and Actinide Elements

Most of the reported photochemical reactions of lanthanide complexes involve some type of redox behaviour. Photolysis (254–405 nm) of Eu^{2+} in acidic aqueous solution, for example, results in Photooxidation of the metal and generation of H_2. While the excited states responsible for this reaction nominally arise from 4f $\longrightarrow$ 5d transitions localized on the metal, strong mixing of the 5d–orbital with ligand orbitals endows these states with appreciable CTTL character. Photoreduction of aquated Eu^{3+} can also be driven with UV (≤ 254 nm) light and forms the basis of a photochemical separation of europium from other trivalent lanthanide elements.

The primary photoreaction, summarized, is consistent with the population of a CTTM state. Several photoredox processes involving the Ce^{4+}/Ce^{3+} couple have been reported. Examples include bimolecular electron transfer from the 4 f $\longrightarrow$ 5d excited state of Ce^{3+} to Cu^{2+} in aqueous solution and the Ce^{4+} mediated oxidation of organic substrates such as alcohols, acids, aldehydes and ketones.

$$Eu^{2+}{}_{(aq)} + H^+ \xrightarrow{hv} Eu^{3+}{}_{(aq)} + \frac{1}{2} H_2 \quad ...(1)$$

$$Eu^{3+}{}_{(aq)} + H_2O \xrightarrow{hv} Eu^{2+}{}_{(aq)} + H^+ + OH \quad ...(2)$$

A few observations of photosubstitution in lanthanide complexes have been reported. Irradiation into the f–f bands of $[Pr(thd)_3]$, $[Eu(thd)_3]$ and [Ho(thd)3] (thd is the anion of 2, 2, 6, 6– tetramethyl–3,5–heptanedione) results in substitution of the by solvent. The proposed mechanism involves intramolecular energy transfer from an f–f excited state to a reactive IL excited state which is responsible for the observed ligand loss. Photosubstitution has also observed upon direct excitation into the ligand absorption bands of $[Tl - c(thd)_3]$.

Knowledge of actinide photochemistry is limited mainly to the complexes of uranium, especially those containing the uranyl ion, UO_2^{2+}. The lowest excited state of this ion is associated with a transition from essentially nonbonding π–orbitals to empty 5 f–orbitals on the U atom. This state has been found to undergo a variety of bimolecular reactions. Upon irradiation in aqueous solutions of primary and secondary alcohols, for example. UO_2^{2+} abstracts a hydrogen atom from the α–carbon atom of the alcohol (equation 3). The resulting Uv species deprotonates and then undergoes disproportionation to UO_2^{2+} and U^{4+}.

Bimolecular electron transfer to photoexcited UO_2^{2+} from a variety of electron donors is now well documented, and a few examples of bimolecular energy donors is now well documented and a few examples of bimolecular energy transfer have also been reported. Complexes of UO_2^{2+} with several carboxylic acids or their conjugate bases possess reactive CTTM excited states which lead to reduction of the metal and oxidation of the organic moiety. The best known example is UO_2^{2}+ oxalate, since this system has found use as a chemical actinometer.

$$UO_2^{2+}{}_{(aq)} + R^1R^2CH(OH) \xrightarrow{hv} UO_2H^{2+}{}_{(aq)} + R^1R^2C(OH)$$

Interest in the reprocessing of spent nuclear fuel has prompted studies of the solution photochemistries of Np and Pu.

PHOTOCHEMICAL REACTIONS OF COORDINATION COMPOUNDS APPLICATIONS

One obvious and immensely important example, is the use of metal-based (particularly silver) systems in photographic processes. Additional applications occur in the area of solar energy conversion and storage. A variety of schemes for photochemical water-splitting make use of transition metal photoredox reactions. Other energy storage processes such as the production of highly strained organic molecules, the photoreduction of carbon dioxide and the generation of electricity in photogalvanic or photo-electrochemical cells also incorporate inorganic photochemical steps. Several applications of inorganic photoreactions are described below :

Synthesis and Catalysis

Photochemical reactions have been exploited to yield coordination compounds that are otherwise difficult to prepare. For example, mixed-ligand complexes of general formula *trans*–$[MClX(en)_2]^+$ (M is Rh or

Ir; X is Br or I) can be obtained in high yields by initial photochemical conversion of aqueous *trans* $[MCl_2(en)_2]^+$ to the corresponding chloroaquo complex as described in equation (1). Precipitation of free Cl^- with Ag^+ followed by heating with one equivalent of X^- then yields the desired product. Thermal routes to these complexes are less desirable because of the propensity to form disubstituted products and the need for extended reaction times at high temperature. Photosubstitution reactions are also employed in the syntheses of a large number of $[RuX(bipy)_2L]^+$ and $[RuX_2(bipy)]$ complexes, where L is an uncharged ligand such as py or CH_3CN and X is an anion such as ClO_4, NO_3, Br or Cl.

$$trans[MCl_2(en)_2]^+ \xrightarrow{hv} trans[MCl(en)_2(OH_2)^2 + Cl^-$$

Chemical Actinometry

The accurate determination of incident light intensity is of pivotal importance in any quantitative photochemical experiment. While various physical devices are available for making absolute intensity measurements, these devices can be difficult to calibrate and usually are rather expensive. A much simpler approach involves the use of a chemical actinometer. This type of system is based upon a photochemical reaction for which product quantum yields are reasonably insensitive to variations in reactant concentration, temperature, light intensity and excitation wavelength. Once the quantum yield is calibrated by an absolute method, a chemical actinometer becomes a rapid, inexpensive and highly accurate secondary standard for light intensity measurements.

Most solution-phase actinometers in use today are based upon the photochemical reactions of coordination compounds. The redox decomposition of acidic solutions of $[Fe(C_2O_4)_3]^{3-}$, for example, is very sensitive to light in the range 254–500 nm. A very good actinometer for longer wavelengths (316–750 nm) involves the photoaquation of *trans*–$[Cr(NCS)_4(NH_3)^{2-}$ (the anion of Reinecke's salt). The extent of reaction in these systems is determined spectrally following the addition of a reagent which forms a highly absorbing complex with one of the photoproducts.

A chemical actinometer especially designed for intensity measurements on high power lasers in the 330–520 nm range has been described. It consists of a closed O_{2-} filled system containing methanolic $[Ru(bipy)_3]^{2+}$ and tetramethylethylene (TME). Photoexcited $[Ru(bipy)_3]^{2+}$ is quenched efficiently by dissolved O_2 to generate singlet oxygen which ten reacts with TME to form a nonvolatile hydroperoxide. The laser intensity is determined from the rate of O_2 consumption monitored on a gas buret.

Finally, the photoredox decomposition of $K_3[Mn(C_2O_4)_3].3H_2O$ has been proposed for use as an actionometer for solid state photochemical studies. Upon UV irradiation, the crystals of this complex change from deep red to white and the extent of reaction can be measured easily by reflectance spectroscopy.

Photochromism

Photochromism may be defined as the reversible light-induced interconversion of a system between two forms having distinguishably different absorption spectra. The generalized process is outlined in equation (1) where ϕ represents the quantum yield for photochemical reaction and k denotes the rate constant for the corresponding thermal process, the system attains a photostationary state. In the case where $\phi_A \geq \phi_B$ and $k_B \geq k_A$, only the photochemical forward reaction and thermal reversion need to be considered and the system is termed singly photochromic.

$$A \underset{\phi B \cdot kB}{\overset{\phi A \cdot kA}{\rightleftharpoons}} B \qquad ...(1)$$

The typical response of a singly photochromic system to light is depicted in Fig. 6.15. Initially, only from A is present, but upon exposure of the system to irradiation at time t_1, the photochemical conversion of A to B ensues. The B/A ratio thus, increases until it attains the equilibrium value of the photostationary state. Upon removal of the light source at time t_2, the photoreaction ceases and B reverts to A at a rate characteristic of the thermal back reaction. The potential commercial and military applications of such reversible light-sensitive materials have provided much of the impetus for research on photochromic behaviour.

A partial listing of applications includes chemical switches, data displays, information storage, radiation intensity control (*e.g.*, sunglasses) and camouflage. The majority of inorganic systems reported to exhibit photochromism are solids, examples being alkali and alkaline earth halides and oxides, titanates, mercuric chloride and silver halides. The coloration is generally believed to result from the trapping of electrons of holes by crystal lattice defects. Alternatively, if the sample crystal is doped with an impurity capable of existing in variable oxidation states (*i.e.*, iron or molybdenum), an electron transfer mechanism is possible.

In solution the photochromism of coordination compounds can result from photoinduced isomerization (*e.g.*, metal dithizonates), substitution (*e.g.*, iron (II) phthalocyanine in the presence of benzyl isocyanide) and oxidation-reduction (*e.g.*, tris(diethyldithiocarbamato)-nickel (IV) bromide).

PHOTOCALORIMETRY

Photocalorimetry is a technique for determining the ordinary enthalpy (ΔH) of a reaction but, unlike conventional calorimetry, the reaction is light induced. Essentially, the procedure involves measuring the rates of heat production in two irradiated solutions, one containing an absorbing but unreactive substance and the other containing the photosensitive compound. The difference between these rates, per mole of reaction, gives the ΔH for the photochemical process.

Photocalorimetry offers a convenient alternative to other methods of ΔH determination and, in some instances, may be the only practical method. The ligand substitution reactions of robust Werner-type complexes are a case in point. Conventional thermochemical measurements are complicated by the slowness of the substitution process and/or by competing reactions. Some these same complexes, however, undergo clean photosubstitutions with high quantum yields and thus, are excellent candidates for photocalorimetry. Examples include $[Cr(NH_3)_6]^{3+}$, $[Cr(CN)_6]^{3-}$ and $[Co(CN)_6]^{3-}$. Photocalorimetric measurements of ΔH have also been obtained for isomerization and redox reactions of co-ordination compounds.

Photoelectrolytic Cell

When an n–type semiconductor is immersed in a solution containing a suitable redox couple A+/A, a few electrons near the surface of the semiconductor react with the oxidizing agent of the couple, A^+, and thus, deplete the semiconductor surface region of charge carriers and cause a bending of the bands.

Absorption at the interface of a photon with energy greater than the band gap promotes an electron to the conduction band and creates a hole in the valence band. Because of the band bending, the electron and hole spontaneously separate, the electron e^- moves toward the bulk of the semiconductor, and the hole (h^+) remains at the interface. If the reduction potential of the solution is low enough, the electron-transfer reaction

$$A + h^+ \longrightarrow A^+$$

occurs at the interface. The excited electron can move through a wire connected to the semiconductor to a nonphotoactive electrode where an oxidized species. B^+ can be reduced.

$$B^+ + e^- \longrightarrow B$$

This combination of n–type semiconductor and inert electrode immersed in the solution comprises a photoelectrochemical cell in which light promotes the reaction.

$$\mathbf{A + B^+ \longrightarrow A^+ + B}$$

If the A/A^+ and B^+/B couple are H_2O/O_2 and H^+/H_2 couples.

$$H_2O \rightleftharpoons \frac{1}{2} O_2 + 2H^+ + 2e^-$$

$$2H^+ + 2e^- \rightleftharpoons H_2O$$

the net cell reaction will be the photoelectrolysis of water to H_2 and O_2:

$$H_2O \longrightarrow H_2 + \frac{1}{2} O_2$$

In fact, this reaction has been carried out using photoelectrolytic cells involving the semiconductor $SrTiO_3$. Because the band gap of $SrTiO_3$ is about 3.4 V, it is necessary to use ultraviolet light with such cells. The ultraviolet light is converted fairly efficiently (about 25 percent) to storable energy in the form of hydrogen and oxygen. Unfortunately, because sunlight consists mainly of visible light, the device has very low efficiency (about 1 percent) for converting solar energy into hydrogen and oxygen. Practical solar photoelectrolysis requires the development of semiconductors that can use visible light efficiently and that have long-term stability under photoelectrolysis conditions. Some fluorine-doped oxide semiconductors show promise for such applications. Thus, $TiO_{2-x}F_x$ (where $x \approx 0.01$) are n–type semiconductors which have photoelectrolytic properties superior to those of the corresponding defect semiconductors, TiO_{2-x} and WO_{3-x}

The concepts learned from photoelectrolytic cells with semiconductor electrodes have been used to design relatively economical systems in which aqueous suspensions of semicounductor particles are used for similar purposes, In a typical system, n–type $SrTiO_3$ powder with platinum dispersed on the surface of the particles has been used. Each particle is essentially a short-circuited photoelectrolytic cell in which the semiconductor electrode and counter-electrode have been brought into contact. Irradiation of such particle suspensions still involves the electron/hole formation and surface oxidation and reduction reactions found in cells, but without external current flow. The Pt–$SrTiO_3$ system, when irradiated with ultraviolet light, has been shown to effect the decomposition of water into hydrogen and oxygen, the hydrogen forming at the platinum surface and the oxygen forming at the $SrTiO_3$ surface. Unfortunately, just as in the corresponding photoelectrolytic cell process, the system has a low solar efficiency.

Solar Energy Conversion and Storage

The photosynthetic membrane in green plant resembles a semiconductor electrode since it can separate the positive and negative charge carriers. This concept is supported by direct measurement on intact chloroplasts which demonstrate the Hall effect on illumination in presence of a magnetic field. Two different charge carriers are produced, one positive and the other n negative, having different lifetimes and different mobilities. The capacity of such membranes to convert solar energy into chemical energy in an endoenergetic reaction ($\Delta G > 0$) in photosynthetic cycle has stimulated interest in development of such model systems for storage and utilization of solar energy by direct conversion of light quanta. In photosynthesis solar energy is stored in two products, carbohydrate and oxygen. The energy is released when carbohydrates burn in air.

Solar energy quantum conversion devices are based on two objectives

1. Conversion of light energy into thermal energy by suitable energy storing *i.e.*, endergonic photochemical reactions,
2. Conversion of light energy into electrical energy by suitable photoelectrochemical devices.

The energy photoreactions occur with positive free energy change ($+\Delta G$) and are hence, thermodynamically unstable.

If the reverse back reaction is prevented or is forbidden by other considerations, the energy remains stored in the photoproducts. Some simple photorearrangement reactions which are governed by Woodward-Hoffman rules have been found usefil.

These rules provide the stereochemical course of photochemical rearrangement based on symmetry properties of the highest occupied molecular orbital (HOMO) and lowest unoccupied molecular orbital (LUMO) or the molecule. A reaction which is photochemically allowed may be thermally forbidden. From the principle of microscopic reversibility, the same will be true for the reverse reaction also. Thermally forbidden back reaction will produce stable photoproducts.

A reaction which can store about 1092 J per gram of material and has shown promise is valence isomerization of norbornadiene (NBD) to quadricyclene. The disadvantage is that it does not absorb in the visible region and therefore the sunlight effiency is poor. By attachment of chromophoric groups or by use of suitable sensitizers it might be possible to shift the absorption region towards the visible for better utilization of solar energy. Photodecomposition reactions in which photoproducts can store energy separately as fuels have been found to be more efficient systems. The photoproducts can be mixed at a later time to regain the stored energy as in the combustion of carbohydrates. If the products are gases, the problem is much simplified although some loss of free energy occurs due to entropy increase. The stable photodecomposition of NOCl occurs with high occurs with quantum yield in the following two steps storing positive free energy $\Delta G^o = +40$ kJ.

$$NOCl + h\nu \longrightarrow NO + .Cl \qquad \Delta G^o = +40 \text{ kJ}$$

$$.Cl + NOCl \longrightarrow NO + Cl_2$$

The products can be made to recombine in a fuel cell.

$$NO + 1/2\ Cl_2 \longrightarrow NOCl \qquad \Delta E^o = 0.21 \text{ V}$$

$$i_{sc} = 8.1 \times 10^3 \text{ A cm}^{-2}$$

(short circuit current)'

Most interesting and fairly successful experiment is the photosensitized decomposition of water. The decomposition of water to H_2 and O_2 is a highly endothermic process

$$H_2O(l) \longrightarrow H_2 + \frac{1}{2} O_2$$

$$DH^o = +295 \text{ kJ}\Delta$$

$$DG^o = +237 \text{kJ}$$

Hydrogen is a nonpolluting fuel which burns in air to produce water again releasing large amount of heat. Alternatively, H_2 and O_2 may be made to recombine in a fuel cell generating electrical energy.

The thermodynamic breakdown energy for water is 1.23 eV. The electrechemical decomposition of water requires two electrons in consecutive steps.

At the cathode $2H_2O + 2e^- \longrightarrow H_2 + H_2O$

At the anode $2OH^- \longrightarrow O_2 + 2H^+ 4e^-$

Therefore is should b e possible to decompose water with two quantum photochemical process with the input of at least 237 kJ or 2.46 eV per molecule. For a one quantum process, the light of wavelength shorter than 500 nm can only be effective.

Which means poor utilization of solar spectrum. Semiconductor electrodes with suitable band gaps can act as quantum collectors and on irradiation with energy E greater than the bad gap E_g can cause electrochemical oxidation or reduction of species in solution. On photolysis of water (0.1NNaOH solution) using n–type TiO_2 semiconductor electrode as the anode and a platinised platinum electrode as the cathode, O_2 is evolved at the TiO_2 electrode on irrdiation. At the dark electrode, H_2 is evolved.

It is necessary to exclude atmospheric oxygen otherwise it will reduced preferentially at the cathode. The phenomenon was first observed by Fujishima and Honda. The steps involved in photoelectrolysis of water at the two electrodes are :

$$TiO_2 + 2hv \longrightarrow 2e^- + 2h^+ \text{ (hole)}$$

$$2h^+ + H_2O \longrightarrow 1/2\ O_2 + 2H^+ \quad \text{(at } TiO_2 \text{ electrode)}$$

$$2e^- + 2H^+ \longrightarrow H_2 \quad \text{(at Pt electrode)}$$

Hydrogen and oxygen can be collected separately and combined again in a fuel cell to obtain electrical energy.

Photoelectrochemical devices to convert light energy to electrical energy implies construction of a battery which undergoes cyclical charging and discharging processes. On illumination the cell is charged.

$$\underset{oxd_1}{A_1} + \underset{red_2}{D_2} \underset{\Delta}{\overset{hv}{\rightleftharpoons}} \underset{red_1}{D_1} + \underset{oxd_2}{A_2}$$

Light energy is converted into chemical energy by driving a suitable redox reaction against the potential gradient where D and A represent the reduced forms and the oxidized forms and the oxidized forms of the donor-acceptor system respectively. The reaction reverses spontaneously in the dark using the external circuit for electron transfer.

In the process chemical energy is converted into electrical energy and the cell is discharged. No consumption of chemical occurs during the charge-discharge cycle. Therefore in an ideal system, the system should function as a rechargeable battery without any loss of efficiency. These are known as *photogalvanic cells.* An example of such a reaction is the reversible photobleaching of thionine by Fe^{2+} ion. Thionine exists as a cation at low pH and can be represented as TH^+. Overall reaction is

$$TH^+ + 2Fe^{2+} + 2H^+ \underset{dark}{\overset{h\nu}{\rightleftharpoons}} TH^+_3 + 2Fe^{3+}$$

The sequence of reaction at the two electrodes is :

$$TH^+ (S_0) + h\nu\ (\lambda max \simeq 600\ nm) \longrightarrow TH^+ (S_1)$$

$$H^+ + TH^+ (S_1) \longrightarrow TH_2^{2+} (T_1)$$

$$TH_2^{2+} (T_1) + Fe^{2+} \longrightarrow .TH_2^+ + Fe^{3+}$$

$$2TH^+_{2.} \longrightarrow TH^+. + TH_3^+$$

$$TH_3^+ + Fe^{3+} \longrightarrow TH_2^+ + Fe^{2+} + H^+$$

$$TH_2\ ^+ \longrightarrow TH_2^+ + H^+ + e^- \text{ (illuminated electrode)}$$

$$Fe^{3+} + e^- \longrightarrow Fe^{2+} \text{ (dark electrode)}$$

$$TH_{2.}\ ^+ + Fe^{3+} \longrightarrow TH^+ + H^+ + Fe^{2+}$$

$TH^+ (S_1)$ = singlet state of dye cation ; $TH_2^{2+} (T_1)$ = triplet state of protonated dye; Th_2^+ = semithionine half-reduced radical ion, TH_3^+ = leuco thionine.

The efficiency of photogalvanic devices for solar energy conversion is not very good mainly because of difficulties in preventing the spontaneous back preventing the spontaneous back recombination reactions. Furthermore they cannot utilize the total solar spectrum. Until now silicon solar cells have been found to be the most efficient quantum conversion devices. They are photovoltaic cells in which light energy is directly converted into electrical energy without the intermediate chemical reaction. The photocurrent is used to charge a storage battery. Since silicon of very high purity is required, they are fairly expensive.From the point of view of light stability and range of absorptivity, inorganic redox systems might be more interesting.

Photoinduced electron transfer in an aqueous solution of tris–(2,2′–bipyridine) ruthenium(II) has been found to decompose water in to a mixture of H_2 and O_2. The complex can serve both as an electron donor and electron acceptor in the excited state. The efficiency is low because of barrier to electron transfer. When spread as a monolayer on glass slides after attaching to a surfactant follow the Stern-Volmer equation, from which the lifetime of triplet molecules τ_D can be calculated.

When these rare earth ions are complexed with suitable organic ligands, such as the diketone-1, 3–propanedione, energy absorbed by the ligand appears as the line emission of the central metal ions, if the triplet state of the ligand lies above the emitting energy level of the ion. The most probable path of energy transfer within the chelate is :

(i) $L_{S1} \longrightarrow L_{T1} \longrightarrow M^* \longrightarrow M_0$

L_{S1}	L_{T1}	M^*	M_0
ligand	ligand	metal	metal
singlet	triplet	excited	ground

If the ligand triplet lies more than 8.4 – 12.6 kJ below the emitting level of the ions, energy is not transferred and broad molecular phosphorescence characteristic of the ligand is observed. Few other schemes are also visualized depending on the relative energy levels of ligand and the metal ions, such as :

Energy level scheme incorporating probable pathways for energy transfer within the chelate is given in Figure 6.24. In these chelates, transfer occurs by exchange mechanism. If suitable couplin interaction exists, transfer may occur from the singlet state of the ligand to the higher energy state of metal ion. The rate of energy transfer $k_{ET} \simeq 10_{10} - 10_{12}$ s^{-1} is predicted.

Energy Transfer Processes Involving Coordination Compounds

Energy transfer from singlet or triplet states of suitable organic molecules can cause excitation of the central metal ion in a coordination compound. Photosensitization of Cr(III) complexes by biacetyl leads to aquation reaction of $Cr(NH_3)_5(NCS)^{2+}$ ion. The aquation of (NH_3) is hundred times more than that of (NCS) for the sensitized reaction, whereas it is only 66 times on direct excitation of Cr(III). This shows that energy transfer occurs from biacetyl triplet to the quartet state of Cr(III) which favours photoaquation of ammonia.

In hexacyaniochromate (III) ion, the lowest doublet and quartet excited states are well separated. By the use of suitable sensitizers, it is possible to excite the two sates separately. By such techniques, it has been established that photoaquation of cyanide ion occurs from the excited quartet state $^4T_{2g}$ whereas ^2E_g state is photochemically inactive and the seat of phosphorescence emission. The intersystem quartet to doublet crossing is not always fast. Similar sensitized aquation and redox reactions are observed for Co (III). Pt (II), and Ru (II) complexes.

MIGRATION OF EXCITATION ENERGY

Exciton Transfer

At very close separation of the donor and acceptor molecules we have seen that Forester type pure dipole-dipole reasonance transfer can become very fast. In the limiting case of a crystal or a very concentrated solution, the transfer rate is so high that the excitation can no longer be regarded as associated with any particular molecule at a particular time and one speaks of an *excition process or delocalized excitation.* In solution the process can be kinetically treated in terms of random walk of the excitation. The transfer rates are proportional to R^{-3}.

In systems where the molecules are stacked in defined geometry such as molecular aggregates, molecular crystals or linear polymers, a set of degenerate energy levels is obtained in which any one of the N molecules may be considered to be excited. The interaction between the transition dipoles of these zeroth order states causes the splitting of the energy levels (Davydov splitting). For example, for a dimer AA, the excited state can be described as AA* and A*A. The wave function for these two descriptions re degenerate. In-phase and out-of-phase interaction of transition dipole M, created in either molecule, destroys the degeneracy. A set of two energy levels results. The interaction is highly geometry dependent and the energy difference ΔE between the two excition levels is given as

$$\Delta E = \frac{2|M|^2}{R^3} G$$

where R is the distance between the centres of the two molecules and G defines the geometry of stacking. The interaction energy varies between 200 cm^{-1} for a strong coupling case and 10 cm^{-1} for a weak coupling case. Although all the excition levels are not equally allowed, they are important for energy migration of excition delocalization. Using time dependent perturbation equation, time period for transfer τ_{tr} is given as

$$\tau_{tr} = \frac{h}{2\Delta E} = \simeq 10^{14}\ s^{-1}\ (\text{for } \Delta E \simeq 1500\ cm^{-1})$$

the fastest rate predicted by any mechanism. Even when the splitting is about 10 cm^{-1} the rate of excition migration is fairly high and much greater than that calculated for diffusion-controlled rate. A well known example of such fast excition migration is a mixed crystal of anthracene containing very small concentration of naphthacene $\simeq 10^{-4}$ ppm as trace impurity. When excited in anthracene absorption band, emission from naphthacene is observed with high efficiency. The host-sensitized emission from guest molecules suggests that the excitation energy explores about 10^6 lattice sites during the lifetime of the excited state. There is no transfer if these crystals are dissolved in fluid or rigid solutions. Such excition transfer modes are said to work in photosynthetic units in plants and many other biological systems.

Liquid Scintillators

If a very dilute solution of anthracene or p–terphenyl is excited by X–rays or γ–rays, emission from the solute present in trace amount is observed. In such systems excition energy supplied by the ionising radiation preferentially interacts with the solvent and finally resides in the low lying excited singlet state of the solvent molecule. Because of the close proximity of like molecules excitation energy migrates from solvent to solvent until it is captured by a solute molecule.

This molecule then becomes the centre of emission with high efficiency. Such solutes are called liquid scintillators and are widely used for detection of ionizing radiations. Sometimes they are embedded plastics or other polymeric meterials. The important scintillator solutes are oxazole and oxazole derivatives. The scintillator diphenyloxazole is commonly called PPO. For liquid scintillators, the nature of solvent is also important. Good scintillation solvents are generally unsaturated molecules such as benzene, toluene and xylene chain, visible evolution of gases occur when a battery of about 20 to 30 slides are exposed to sunlight.

Many reducing cations in aqueous solutions can photoreduce water in low yields when irradiated in their charge transfer bands. Similarly, oxidizing cations can photo-oxidize water. Several hydride complexes of transition metals such as cobalt decompose on irradiation to yield molecular hydrogen.

$$CoL_4H_2 \xrightarrow{hv} CoL_4 + H_2$$

The complexes with electron withdrawing ligand (L′) on the other hand can decompose water spontaneously

$$CoL_4' + H_2O \xrightarrow{h\nu} CoL_4'\,H_2 + 1/2\ O_2$$

If suitable system of ligands could be found such that light reaction is endergonic and dark reaction is exergonic, then the net result will be decomposition of water.

The abiological fixation of CO_2 and H_2O to reduced compounds which can be used as fuels will be a reaction of immense use. Unit now this has not been very successful. The simplest of these compounds are formic acid, formaldehyde, methanol and methane :

$$CO_2 + H_2O \longrightarrow HCOOH + 1/2\ O_2; \quad \Delta H^0 = 262\ kJ\ mol_{-1}$$

$$CO_2 + H_2O \longrightarrow HCHO + O_2; \quad \Delta H^0 = 560\ kJ\ mol^{-1}$$

$$CO_2 + 2H_2O \longrightarrow CH_3OH + (2/3)\ O_2; \quad \Delta H^0 = 713\ kJ\ mol^{-1}$$

$$CO_2 + 2H_2O \longrightarrow CH_4 + 2\ O_2; \quad \Delta H^0 = 690\ kJ\ mol^{-1}$$

All these reaction are highly endothermic. The products of these reactions are high calorific fuels. In nature fixation of CO_2 is a dark process although initiated by light induced chain of reactions. Except in some special plants like rubber, CO_2 is still not completely reduced. Carbohydrates are the common stored products. Rubber plants produce hydrocarbons.

MECHANISMS

Photochemical reaction mechanisms are deduced from measurements of the dependence of quantum yields on such different reaction variables as reactant concentration and the concentration of added substances. Primarily it is necessary to identify the nature of the electronically excited state responsible for of the reaction, which is most often singlet ($^1M^*$) or triplet) ($^3M^*$) or triplet ($^3M^*$) following which a theoretical examination of the orbital transformations along the reaction coordinate may be attempted.

If the quantum yield is reduced by a selective triplet-state quencher (Q) which introduces the competing process reaction then the triplet state $^3M^*$ may be regarded as the reactive state; otherwise

$$^3M^* + Q \longrightarrow M + {^3Q^*} \qquad ...(1)$$

the singlet state $^1M^*$ is assigned the role of reactive intermediate. The rate constant of the primary process may then be obtained by monitoring the decay of intermediate [1]. $^3M^*$ in absorption or emission following flash of laser pulse excitation. Photochemical reactions are classified as unimolecular, in which the excited molecule itself undergoes chemical change, or bimolecular if the excited molecule reacts with another molecule present.

Quantum Yield

The rate of a photochemical reaction is proportional to the rate at which electronically excited molecules M* are produced ; Since each absorbed photon (energy hV) of light produces one excited molecule, this excitation rate is equal to the rate of light absorption or the intensity of light absorbed I_a. The reaction rate is therefore proportional to Ia or rate = γ/I_a, where the proportionality constant, γ = rate/I_a, known as the quantum yield (or efficiency) is characteristic of the reaction under the conditions of examination. The quantum yield of the primary process (that involving the excited molecule) cannot exceed unity and is usually much lower than this since electronically excited molecules revert to the ground state (with or without the emission of luminescence) within a period of ~ 10^{-3}s (triplet state) to 10^{-s} s (singlet states).

These so-called photophysical processes, summarized in the illustration, complete with photochemical reactions, effectively reducing the quantum yield. Additionally, the excited molecule may be quenched by other molecules Q in a process of electron transfer [reaction (2)] or energy transfer [reaction (3)], which leave the molecule M chemically unchanged.

$$M^* + Q \longrightarrow (M + Q^-)$$

or $$(M^- Q^+) \longrightarrow M + Q \quad ...(2)$$

$$M^* + Q \longrightarrow M + Q^* \quad ...(3)$$

On the other hand, if the primary photochemical products are atoms or free radicals, these may undergo secondary non-photochemical chain reactions which amplify the primary process and lead to overall photochemical quantum yields much greater than unity. An example is reaction (3), in which the primary process (4b) is followed by the chain propagating steps (4c and 4d) to produce ~10^6 molecules of HCl for each light photon absorbed by Cl_2, that is, an overall quantum yield of 10^6.

$$H_2 + Cl_2 \xrightarrow{\text{light}} 2HCl \quad ...(4a)$$

$$Cl_2 + \text{light} \longrightarrow 2Cl \quad ...(4b)$$

$$Cl + H_2 \longrightarrow H + HCl \qquad ...(4c)$$

$$H + Cl_2 \longrightarrow Cl + HCl \qquad ...(4d)$$

PHOTOCHEMISTRY

It deals with the study of the light-induced behaviour of various metal compounds. The physical and chemical properties of substances are generally altered by the absorption of light. Typical metal compound have a characteristic number (coordination number) of molecules or ions (ligands) directly bonded to the metal centre. For example, a six-coordinate compound has the general formula ML_6^{n+}. Many of these compounds are coloured, and much interest has been aroused by speculation that some metal compounds could mediate the transformation of solar radiation into useful chemical or electrical energy.

The photochemistry of metal compounds has grown in concert with modern theories of the electronic structure of molecules and of chemical bonding in molecules. Photochemical studies are often designed to probe and test these theories. The range of pertinent studies spans most of the subdisciplines of chemistry and includes or bears on such topics as photophysics, the development of laser materials, catalysis, photosynthesis, oxidation-reduction chemistry, acid-base chemistry, organometallic chemistry, metalloenzyme chemistry, solid-state chemistry, and surface chemistry.

In *Inorganic photochemistry*, an area of rapidly growing interest in inorganic chemistry, energy is imparted to molecules by irradiating them with visible and ultraviolet light. Energy of light quanta ranges from ~ 170 to 590 kj/mol-up to 3.5 times that available thermally at ordinary temperatures. Only certain wavelengths of light are absorbed by transition metal complexes-namely, those corresponding to difference between electronic energy states.

The photochemical excited state may luminesce with a rate constant k_r, undergo internal conversion with a rate constant k_q, or react (perhaps with other species present) with rate constants k_a, k_b,... and so on. The figure also indicates schematically how products thermodynamically (A) or kinetically inaccessible (C) in thermal reactions might result from photochemical excitation. Four types of photochemically excited states can be achieved for transition metal complexes. Ligand-field excited states, which result from d–d transitions and generally place electron density in $e_g\sigma^*$ orbitals, are expected to lead to weakened L–M bonds and to favour ligand labilization. Charge-transfer (CT) excited states may be L →M or M → L. IN the latter, more common, case, the reactivity of the excited state should parallel that of L or involve electron transfer to the oxidized metal centre. Transitions localized on the ligands result in intraligand excited states.

Insofar as anything is known about the reactivity to be expected, ligand rearrangement and ligand reaction with other substrates may be involved. Finally, a charge transfer to solvent state involving an oxidized metal ion and a solvated e_s could lead to electron transfer to the oxidized metal and reduction of solvent or other substrate by e_s. As one example, photoaquation of $[Co(NH_3)_6]^{3+}$ to give $[CO(NH_3)_5(H_2O)]^{3+}$ occurs readily upon irradiating the lower-wavelength ligand-field band of the hexammine. In contrast, an acid solution of $[CO(NH_3)_6]^{3+}$ can be boiled for days without a reaction occurring. The photochemically excited state is~ 10^{13} as reactive as the ground state.

Irradiation of $[Ru(bipy)_3]^{2+}$ and related complexes produces a CT excited species, $[Ru^*(bipy)_3]^{2+}$, with an oxidized Ru and an electron localized on bipy. This excited molecule is both a better oxidizing agent and a better reducing agent than $[Ru(bipy)_3]^{2+}$, as the reduction potentials in Table 2 show. $[Ru^*(bipy)_3]^{2+}$, with an oxidized Ru and an electron localized on bipy. This excited molecule is both a better oxidizing agent is luminescent : quenching of the luminescence by other species in solution Q is evidence of a reaction that may be oxidation, reduction, or energy transfer.

$$[Ru^* (bipy)_3]^{2+} + Q \longrightarrow [Ru(bipy)_3]^{3+} + Q^- \quad ...(5)$$

$$[Ru^* (bipy)_3]^{2+} + Q \longrightarrow [Ru(bipy)_3]^{1+} + Q^+ \quad ...(6)$$

$$[Ru^* (bipy)_3]^{2+} + Q \longrightarrow [Ru(bipy)_3]^{2+} + Q^* \quad ...(7)$$

Table 2 : Reduction Potentials for some Ru Complexes at 25°C.

Reaction			*E^0(Volts)*
$[Ru(bipy)_3]^{3+}$	$+ e$	$\longrightarrow$	$[Ru(bipy_3]^{2+}$ + 1.26
$[Ru^*(bipy)_3]^{2+}$	$+ e$	$\longrightarrow$	$[Ru(bipy_3]^{1+}$ + 0.84
$[Ru(bipy)_3]^{3+}$	$+ e$	$\longrightarrow$	$[Ru^*(bipy_3]^{2+}$ + 0.84
$[Ru(bipy)_3]^{2+}$	$+ e$	$\longrightarrow$	$[Ru(bipy_3]^{1m+}$ + 1.28

The oxidation and reduction reactions represent way of converting light energy into chemical energy. Energy storage could also be achieved if a suitable quencher could convert the excited species into complexes such as the excellent oxidizing agent $[Ru(bipy)_3]^{3+}$ and the excellent reducing agent $[Ru(bipy)_3]^{+}$. (The first of these has been shown to oxidize OH– to O_{2-} representing a possible step in the decomposition of water to H_2 and O_2 by solar energy. One such set of thermodynamically feasible quenching reactions is

$$[Ru^*(bipy)_3]^{2+} + [Ru(NH_3)_6]^{2+} \longrightarrow [Ru(bipy_3)^{3+} + [Ru(NH_3)_6]^{+} \quad ...(8)$$

$$[Ru^*(bipy)_3]^{2+} + [Ru(NH_3)_6]^{2+} \longrightarrow [Ru(bipy_3)^{1+} + [Ru(NH_3)_6]3^{+} \quad ...(9)$$

Unfortunately, both thermodynamics and kinetics favour the back-reaction of the products to regenerate $[Ru(NH_3)_6]^{2+}$, dissipating the stored energy to the solution. A possible solution for this problem stored as either $[Ru(bippy)_3]^{3+}$ or $[Ru(bipy)_3]^{1+}$. Gray and co-workers have prepared isocyanide complexes such as $[Rh_2\ (\mu_2\text{–}NCCH_2CH_2CN)_4]^{2+}$, which on irradiation reduces H^+ to H_2. Both kinds of complexes have potential application to solar energy storage and splitting of water. However, our main focus here is on the fact that photolytic activation can lead to chemistry quite different from that caused by thermal activation.

An interesting connection exists between photochemistry and the redox reactions just discussed. Taube, Creutz, Meyer, and others have prepared a series of mixed-valence compounds of which $[(bipy)_2ClRU^{11}$–

N 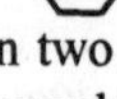 –N–$Ru^{11}(bipy)_2Cl]^{3+}$ is an example. These contain the same metal in two different oxidation states and serve as models for the precursor complex in inner-sphere e transfer reactions when the bridging ligand contains some unsaturation. We can study the rates of electron transfer apart from considerations relating to the precursor stability.

Mixed-valence complexes usually are intensely coloured because of an absorption in the low-energy region. This absorption is not present in the spectrum of $[Ru(bipy)_2Cl(py)]^{n+}$ (n = 2,3), and results from electron transfer from one metal to the other-the so-called *intervalence transfer* (IT). This figure is like in that the reactants and products are identical. However, E_{op} (the IT energy) differs from E_{th} (the thermal energy for e transfer) because of the Franck-Condon principle. Most of the complex ions in solution are at the equilibrium position for the ground state when the photon energy is absorbed. Because nuclear motions are slow compared with electronic transition, the energy absorbed in going to the upper state is E_{op}, corresponding to a vertical transition. For symmetrical complexes, we can show that $E_{op} = 4E_{th}$.

Example:

Show that $E_{op} = 4E_{th}$ for a symmetrical mixed-valence complex.

Solution:

Suppose a is the distance between energy minimum and the thermal transition state. Both curves are parabolas. Hence, at the thermal transition state, E_{th}. = K $(a)^2$, where k is the proportionality constant. At the value of reaction coordinate corresponding to the reactants, the distance from the product curve minimum is 2a. Thus, the energy on the upper parabola is $E_{op} = k(2a)^2$. Hence, $E_{op} = 4E_{th}$..

At least theoretically, measurements of E_{op} should lead to estimates of self-exchange in symmetrical complexes. Alternative energy possibilities for unsymmetrical mixed-valence complexes that might contain different metals or different ligands for example, $[Cl(NH_3)_4Ru^{11}$—N  N – $Ru^{11}(bipy)_2Cl]^{3+}$

Unsaturated bridging ligands such as pyrazine give mixed-valence complexes that are good models for inner-sphere precursor complexes, because of orbital overlap between π* ligand orbitals and each separate metal orbital. If the bridging ligand is saturated and sufficiently long, electronic coupling between redox sites is essentially from outer-sphere orbital overlap. Such a complex is $[Cl(bipy)_2Ru^{11}Ph_2CH_2CH_2CH_2PPh_2$ $Ru^{111}Cl(bipyl)_2]^{3+}$, which also exhibits an IT band, but one of much lower intensity, because of unfavourable orbital overlap. Thus, complexes with low-intensity IT bands are models for ourtersphere e-transfer reactions.

Redox chemistry can often be initiated by irradiating the IT. band.

$$[(NH_3)_5Co^{111}N \equiv C-Ru^{11}(CN)_5]^- \xrightarrow{hv} 5\,NH_3 + Co^{2+} + [Ru^{111}(CN)_6]^{3-} \quad ...(10)$$

A third group of complexes also may be prepared from metals in two different oxidation states. Where the bridging ligand has delocalized orbitals and the metal orbitals can overlap with particular effectiveness (because of a short physical separation, low oxidation state, etc.), the two metal centres may "communicate" electronically in the ground state. In these cases, delocalized molecular orbitals exist and the electrons are distributed equivalently over both metals in the ground state. An example is $[(NH_3)_5Ru–O–Ru(NH_3)_5]^{5+}$, where each Ru is assigned 5.5 d electrons. A band corresponding to a transition between two delocalized molecular orbitals now is expected. The delocalized case can be recognized from electrochemical, UV–V is spectral, and magnetic measurements.

Inorganic photochemistry is a comparatively new research area. Important questions demand investigation : what is the exact electronic state that undergoes photoreaction? (This is the photochemists mechanism)? What fraction of the light quanta absorbed leads to a particular reaction? (This is the quantum yield for the reactions and corresponds to the thermal rate constant.) Besides photoreaction, what other processes do excited molecules undergo.

TRANSITION METAL COMPLEXES OF PHOTOCHEMISTRY

The photochemistry of transition metal complexes is more varied than large organic molecules because of the possibility of different orbital types of excitation and of overall symmetry of the complex which may very from O_h to C_s. A set of metal orbitals and a set of ligand orbitals are available. For p–bonding ligands a set of p, p* orbitals are also available. Basically four different types of low lying energy states can be described depending on whether the excitation is localised within the metal ion energy manifold or within the organic ligand energy manifold of delocalized. These are termed as follows :

(i) *d–d states or ligand–field states* : They arise from promotion of an electron from t_{2g} (nonbonding) to e_g (antibonding orbital) for octahedral symmetry, an excitation confined essentially to the metal ion. The energy difference is determined by Δ, the ligand field parameter. The Δ is a function of ligand field strengths, the position of the metal ion in the periodic table and the oxidation state of the ion.

Further splitting of the T states may arise due to interelectronic repulsion terms. These states are labelled according to group symmetry nomenclature. Because of the Laporte forbidden nature, molar extinction values are low, $\varepsilon \simeq 1–150$ $l\,mol^{-1}\,cm^{-1}$. The resulting increase in antibonding electron density decreases the net bonding in the complex and causes a lengthening of the , metal ligand bonds.

(ii) *d–π* states* : These arise form excitation of a metal to a π* antibonding orbital located on the ligand system. This can be considered as transfer of an electron from the metal (M) to the ligand (L) and hence, is termed CTML type. Since such transfers leave the metal ion temporarily in an oxidized state such states are related to the redox potentials of the complex. The (d, π*) state should lie at relatively low energy for easily oxidizable complex. Thus, the change of central metal ion will considerably affect the position of the (d,π*) states. They have charge transfer character and high molar extinction, $\varepsilon \simeq 10^4\ l mol^{-1}$. The reverse process, the charge transfer from the ligand to the metal may also occur, *i.e.,* CTLM type transitions.

(iii) *π, –π* states :* These states arise from localized transition within the ligand energy levels. They lie at relatively high energies. The metal ions perturb them only slightly but can drastically affect established.

(iv) *π – d states :* Such states are expected to arise from a promotion of an electronic charge from ligand π – system to the higher orbitals of the metal (e–type for O_h symmetry). These are not very well established.

Each of these promotional types of energy states can further split by spin-orbital coupling interactions to give singlet and triplet states. For heavier elements the total angular momentum quantum number J becomes a ' good' quantum number. Spin-orbital coupling interaction energy can very from 500 cm^{-1}(Co^{3+}) to 4000 cm^{-1} (Ir^{3+}). The relative of energy levels can be altered by replacing metal ions, exchanging ligands, modifying the ligand or by varying the geometry. These levels can also be modified by solvent effects.

Photophysical Processes

Very few inorganic complexes fluoresce. The classical examples are emission from rare-earths. For example Sm^{3+}, Eu^{3+}. Gd^{3+}, Tb^{3+} when complexed with bidented ligands emissive transitions occur within the metal-energy levels which is formally Laporte forbidden. The levels are populated by intramolecular energy transfer in the complex when radiation is absorbed by the ligand. If the emitting state of the metal does not lie below the lowest triplet state of the ligand moiety, characteristic fluorescence and phosphorescence of the latter may be observed under suitable conditions. Then ion UO^{2+}_2 is also highly luminescent.

Amongst the transition metal complexes, emission is confined to d^3 (Cr^{3+}, Co^{3+}) and d^6 systems (Ru^{2+}, Rh^{3+}, Os^{2+} and Ir^{3+}). These are d–d emitters and spin-forbidden phosphorescence generally predominates. The efficiency ϕ_f is low and is related to the spacing and coupling of lowest excited vibronic levels with the ground state vibronic levels. Flurorescence is generally broad, structured and exhibits large Stoke's shift which indicated geometrical distortion of the excited state. Phosphorescence is characterized by very small stoke's shift and resolved vibrational structure at low temperature in glassy solutions and arises from $^2Eg \rightarrow {}^4A_{2g}$ transitions corresponding to $T_1 \rightarrow S_0$. In these complexes nonradiative internal conversion k_{IC} is relatively more efficient than nonradiative intersystem crossing.

Emission from transition metal complexes obeys Kasha's rule and originate from the lowest excited state which are

(i) $^3(\pi, \pi^*)$ state in $[Rh(phen)_3]$ $(ClO_4)_3$ in water-methanol glass

(ii) $^3(d, \pi^*)$ state in $[Ru(bby)_3]Cl_2$ in ethanol-,ethanol-methanol glass, and

(iii) $^3(d\text{–}d)$ state, in solid $[RhCl_2(phen)_2]$ Cl. Their characteristics differ in details and are given in Figure 6.4. Sometimes weak fluorescence is also observed, from Cr^{3+} complexes. $K_3[Co(CN)_6]$ is highly luminescent. The ϕ_p and τ_p are temperature dependent. Deuterium substitution has large effect on nonradiative processes.

Sometimes $^3(d\text{–}\pi^*)$ and $^3(\pi\text{–}\pi^*)$ states to be derived from delocalized orbitals and $^3(d\text{–}d)$ state from *localized* orbitals. The shift of the chelate emission from that of the free ligand increases in the sequence Rh(III) < Ir (III) < Ru (II) and reflects increasing d–orbital participation in the emission orbital. The decrease in the chelate emission lifetime from the free ligand values also reflect the contamination of the molecular orbitals with d–character. The role of metal complexes as quenchers of excited states of π-electrons in organic compounds can be rationalized from such considerations. Emission from Cr^{3+} is the basis of one of the most important solid state laser system, the Ruby laser.

Photochemical Processes

These reactions involve the absorption of light to produce a highly reactive electronic state. The absorption of a photon increases the energy of a complex, typically by between 170 and 600 kJ mole^{-1}. Because these energies are larger than typical activation energies it should not be surprising that new reaction channels are opened. However, when the high energy of a photon is used to provide the energy of the primary forward reaction, the back reaction is a almost always very favourable, and much of the design of efficient photochemical systems lies in trying to avoid it.

Because of the fast nonradiative deactivation of low lying energy states of transition metal complexes, the activation energy for the reactions that may occur from these states must be zero to enable them to compete effectively. For transition metal complexes both $^4T_{2g}$ and ^2Eg states can be photochemically active but may follow different chemical pathways.

Three fundamental types of photochemical reactions are known for coordination compounds :

(a) Photosubstituion reactions

(b) Photo rearrangement reactions

(c) Redox reactions

Let us discuss these one by one.

PHOTOSUBSTITUTION REACTIONS

Photosubstitution reactions can be equation, anation or ligand exchange.

(i) Photoaquation reactions of the type,

$$Cr^{3+} L_6 + H_2O \xrightarrow{hv} Cr^{3+} L_5(H_2O) + L$$

involve heterolytic cleavage of Cr–L bond. For the specific case of Cr(III) complexes of O_h symmetry, efficient photochemical aquation reaction occurs when the $^4T_{2g}$ state lies significantly above the $2E_g$ state. Efficiency decreases as the energy difference ΔE between these two states decreases. This is happens for $(NH_3)_6{}^{3+}$ on successive aquation. The system deviates from Oh symmetry and consequently ϕ for aquation decreases form 0.3 for hexammine complex to less than 0.002 for Cr $(NH_3)H_2O)_5{}^{3+}$. Relative liabilities of the liabilities of the ligand NH_3 are reversed in the thermal and photochemical reactions.

The $Cr(NH_3)_5\ (NCS)^{2+}$ in 0.1 NH_2SO_4 presents two aquation modes:

$$Cr.(NH_3)_5(H_2O)^{3+} + NCS^- \qquad ...(1)$$

hv

$$Cr(NH_3)_5(NCS)^{2+}+H_2O$$
$$hv$$
$$Cr\ (NH_3)_4(H_2O)(NCS)^{2+} + HNH_3 \quad ...(2)$$

Both $^4T_{2g}$ and $2E_g$ states are photoactive. The ratio of f NH_3/ϕ NCS– is wavelength dependent, the ratios being 15 at 37.3 nm (quarter band); 22 at 652 nm, (doublet band).

The *photosensitization* of this aquation can also be initiated by *energy transfer* from biacetyl and acridinium ion in the same solvent. The energy transfer reactions have helped to identify the details of this aquation reaction:

(a) *Biacetyl as sensitizer :* phosphorescence quenched ; fluorescence not affected, release of NH_3; No apparent degradation of biacetyl; $\phi_{NH3}/\phi_{NCS-} > 100$.

(b) *Acridinium ion as Sensitizer :* Fluorescence quenched; both NH3 and NCS–released although NH3 predominates; in presence of O_2, NCS^- release decreased, hence, triplet state of the sensitizer is involved; $\phi\ NH_3/\phi_{NCS-}$, 33 for quarter state, 8 for doublet state.

Therefore it can be concluded that NH_3 release occurs from $^4T_{2g}$ state and NCS^- from ^2E_g state. $Co(NH_3)_6^{3+}$ is stable in aqueous solution whereas $[Co(NH_3)_5X]^{2+}$ undergoes moderately rapid substitution of water for the acid group X^-.

The rates of aquation reaction depends strongly on basicity of X^- for example, nitrate : acetate is 10^3 : 1. The difference in behaviour between Cr(III) and Co(III) ammines may be due to necessity for energy of activation in the transition state for reaction with the latter.

The photoaquation of $KCr(NH_3)_2(SCN)_4$ (Reinnecke's salt) and $Cr(urea)_6\,Cl_3$ can be used as efficient actinometers between the range 316 and 750 nm. For $Co(CN)_6^{3-}$ the photoreaction

$$Co(CN)_6^{3-} \xrightarrow{hv} Co(CN)_5(H_2O)^{2-} + CN^-$$

is without complications and occurs with high quantum yield $\phi = 0.31$.

(ii) *photoanation reactions :* For example,

$$Cr(H_2O)_6^{3+} + NCS^- \xrightarrow{hv} Cr(H_2O)_5(NCS)^{2+} + H_2O$$
$$Cr(H_2O)_6^{3+} + Cl^- \xrightarrow{hv} Cr(H_2O)_5Cl^{2+} + H_2O$$

(iii) *Ligand photosubstitution Reactions* : These reactions involve the replacement of a coordinated ligand by an incoming ligand from solution. The irradiation of monosubstituted carbonyl $M(CO)_{n-1}$ L in presence of excess of L, causes the formation of bisubstituted complexes: (M = metal, L = ligand). For example,

$$Mo(CO)6 + L \xrightarrow{hv} Mo(CO)5\ L + CO$$

where, L = NH_3, $(C_2H_5)NH_2$, $(CH_3)_2NH$, $(CH_3)_3N$, $(C_2H_5)_3N$ or other ammines. The pentacobaltic complex may undergo further substitution.

Another example is :

$$Fe\ (CN)_6^{4-} + dpy \xrightarrow{hv} Fe(CN)_4(dpy)^{2-} + 2CN^-$$

In the case of mixed ligands photoreactivity may be predicted empirically from considerations of the ligand field strength (Adamson's rule).

For O_h complex, the axis having the weakest average field as determined by the position of its respective ligands in the spectrochemical series will be labilized.

If the labilized axis contains two different ligands, the ligand of greater field strength will preferentially be replaced. For example, upon irradiation of $Cr(NH_3)_5Cl^{2+}$, the NH_3–Cr–Cl axis is labilized rather than NH_{3-} Cr–NH_3.

Furthermore, the stronger field ligand NH_3 is lost along NH_{3-}Cr–Cl axis. Steric and solvation effects may cause deviations from these predictions. Lowering of symmetry of the complex will also affect the nature of reaction due to switching of states.

PHOTOREARRANGEMENT REACTIONS MAY INVOLVE

(i) geometrical photoisomerisation

(ii) photoracemization

(iii) linkage (photoisomerisation)

(iv) ligand (photorearrangement)

(i) Geometrical photo isomerization

$$cis - bis\ \text{(glycinato) Pt (II)} \xrightarrow[\text{band}]{\text{d-d}} trans\text{–isomer}$$

$$\phi \simeq 0.13$$

$$trans \not\longrightarrow cis$$

The photoprocess involves intramolecular twist mechanism without bond cleavage. The reactive intermediate is a triplet state of pseudotetrahedral geometry. The transformation is photochemically allowed but thermally disallowed from symmetry considerations.

(*ii*) *Photoracemization*

The reaction shows deuterium isotope effect.

(iii) Linkage Photoisomerization

$$Co(NH_3)_5(NO_2)^{2+} \xrightarrow{hv(solid)} (NH_3)_5Co(-ONO)^{2+}$$

The nitro-nitrito photoisomerization occurs in solution through an intramolecular mechanism involving the homolytic fission of Co–NO_2 bond and then cage recombination of two fragments by means of a Co–ONO bond. This complex exhibits CT and CTLM character at 239 and 325 nm and a ligand field band at 458 nm.

$$Co(NH_3)_5(NO_2)^{2+} \xrightarrow[\text{in 0.1 NHClO4}]{\text{CT and d-d}} [Co(NH_3)_5{}^{2+}.NO_2]$$

(iv) Photoexchange processes

$$Co(CN)_5(NO)^{3-} + CN^* \xrightarrow{hv} Co(CN^*)_5(NO)^{3-} + CN^-$$

Photo-oxidation Reduction or Redox-reactions

Redox reactions involve changes in oxidation state of the reacting complexes. Inner-sphere redox reactions involve a transient bridging ligand, which carries electron density from one metal to the other. Outer sphere redox reactions involve electron transfer from one reactant species of the other.

A photooxidation-reduction process in solution may be *intramolecular* when the redox reaction occurs between the central metal atom and one of its ligands or *intermolecular* when the redox reaction occurs between the central metal atom and one of its ligands of *intermolecular* when the complex reacts with another species present in the solution.

When central metal ion becomes strongly oxidising on irradiation in the ligand field band (d–d), it can lead to intramolecular reduction of the central metal ion

(*i*) $2Fe^{(IIII)}(C_2O_4)_3{}^{3-} \xrightarrow[0.1\,NH^+]{hv} 2Fe(C_2O_4)_2{}^{2-} + (C_2O_4)^{2-} + 2CO_2$

(*ii*) $Co^{(IIII)}(NH_3)_5\ I^{2+} \xrightarrow{hv} Co^{2+} + 5NH_3 + 1$

The photodecomposition of ferrioxalate in $0.1NH_2SO_4$ (reaction (i) is a very useful chemical actinometer for the near uv and the visible region upto 400 nm ($\phi \simeq 1.12$)– *ferrioxalate* actinometer. In some low-valence hydrated cations, such transition may bring about photo-oxidation:

$$Fe^{2+}.\ H_2O \xrightarrow{hv} Fe^{3+} + OH^- + H$$

Another example is Fe (thionine)$^{2+} \longrightarrow Fe^{3+}$ + semithionine which reverses in the dark. This reversible photoredox reaction has the possibility of conversion of light energy into electrical energy. The irradiation at the general type.

$$(M^{z+}\ L_n)^{z+} \xrightarrow{hv} M^{(z-1)+} + (n-1)L + \text{oxidation products of L.}$$

The excitation leads to homolytic M–L bond fission. Such a reactivity is not expected to d–d band and if redox reaction does occur under such excitation it is likely to hide a low lying reactive CT type band under its envelop.

It is not necessary that the reaction should always lead to decomposition of the ligand. If the central ion can give a stable complex with one lower (or higher) oxidation number, the product of an intramolecular redox reaction may be a complex with a central metal ion of different oxidation number.

$$IrCl_6^{2-} \xrightarrow[H_2O]{hv} IrCl_5(H_2O)^{2-} + CI$$

Intermolecular photo-oxidation-reduction reactions involve a light initiated electron transfer between a complex and any other suitable molecule available in the medium. An oxidized or a reduced form of the complex may be obtained.

$$ML_n^{z+} + X \xrightarrow{hv} ML_n(z+1)^+ + X^-$$

$$ML_n^{z+} + Y \longrightarrow ML_n(z-1)^+ + Y^+$$

If the new complex species is unstable, a decomposition reaction may then follow. Examples :

$$Co(NH_3)_6^{3+} + I^- \xrightarrow{hv} Co(NH_3)_6^{2+} + I$$

$$Mo(CN)_8^{3-} + H_2O \xrightarrow{hv} Mo(CN)_8^{4-} + H^+ + OH$$

In the second example, no decomposition of the complex occurs. The excitation is in the ligand field band. When the new complex is stable, an electron may be ejected to the solvation shell leading to the formation of hydrated electron e^-_{aq}

$$Fe(CN)_6^{4-} + H_2O \xrightarrow{hv} Fe(CN)_6^{3-} + e^-_{aq}$$

This is a CTTS type of transition. The photoejection of electron may also occur if CTLM band increases the charge density towards the periphery of the complex.

(*iii*) *Chemiluminescence in ruthenium complexes.* Ru(III)– 2, 2′ bipyridyl –5 methy –1 –1, 10 –phenanthroline and few other complexes when react with OH^- or hydrazide, generate enough thermal energy to promote it to an electronically excited state. The molecule relaxes by emission of radiation. In such reactions chemical energy is converted into electronic energy. The general reactions scheme is

$$\text{Metal (ligand)}^{(n+1)+} + e^{(-)} \text{ (from reductant)} \rightarrow \text{Metal (ligand)}_x{}^{n+} + hv$$

PHOTOCHEMICAL REACTION

These often involve the rapid dissociation of a ligand upon photoexcitation. These photosubstitution reactions often result from d–d transitions.

Delayed photochemical reactions often involve electron transfer, either ligand-to-metal or metal to-ligand. Long-lived excited states associated with electronic transitions in metal-metal bonded compounds may also lead to delayed photochemical reactions.

In some cases, the excited states dissociate promptly, Examples include formation of the pentacarbonyl intermediates that initiate ligand substitution in carbonyls

$$Cr(CO)_6 \xrightarrow{hv} Cr(CO)_5 + CO$$

and the scission of Co–Cl bonds

$$[Co^{III}Cl(NH_3)_5]^{2+} \xrightarrow{hv(\lambda<350nm)} [CoII(NH_3)_5]^{2+} + Cl$$

Both these processes occur in less than 50 fs and hence, are called prompt reactions. These reactions often involve the rapid dissociation of a ligand upon photo excitation. These photosubstitution reactions often results from d–d transitions. In the second reaction, the *quantum yield*, the amount of reaction per mole of photons absorbed, increases as the wavelength is decreased and the photon energy increases. Energy in excess of the bond energy is available to the newly formed fragments and contributes to the probability that they will escape from each other through the solution before they have an opportunity to recombine.

Some excited states have long lifetimes. They may be regarded as energetic isomers of the ground state that can participate in *delayed reactions*. The excited state of $[Ru^{II}(bipy)_3]^{2+}$ created by photon absorption in the metal-to-ligand charge transfer band may be regarded as a Ru(III) complexed to a radical anion of the ligand. Its redox reactions can be explained by adding the excitation energy (expressed as a potential using $-FE = \Delta G$ and equating ΔG to the molar excitation energy) to the ground state reduction potential :

$$[Ru(bipy)_3]^{3+} + e^- \longrightarrow [Ru(bipy)_3]^{2+}$$

$$E = +\ 1.26\ V\ (-122 kJ\ mol{-}1)$$

$$[Ru(bipy)_3]^{2+} + h\nu\ (590\ nm) \longrightarrow [Ru(bipy)_3]^{2+}$$

$$\Delta G = +\ 20 kJ\ mol^{-1}$$

$$[Ru(bipy)_3]^{3+} + e^- \longrightarrow [Ru(bipy)_3]^{2+}$$

$$E = -\ 0.84\ V\ (+\ 80\ kJ\ mol^{-1})$$

PHOTOCHEMISTRY OF METALLOCENES

Metallocenes act as good quenchers of triplets whose energies lie between 280 to 177 kJ/mol. Quenching rates are nearly diffusion controlled. They are good sensitizers of *cis-trans* isomerization. The diamagnetic ferrocene is light stable to visible and uv but decomposes in presence of O_2 in an acidified solution. In halogenated solvents at room temperature, solutions of ferrocenes are stable in the dark but on exposure to light, decomposition occurs at a relatively rapid rate. In CCl_4, it forms a CT complex. Tetracynoathylene forms a π–type complex with ferrocene (I). Another metallocene is bis–benzenechromium iodide (II) a sandwitch type complex.

d–d CHARGE-TRANSFER REACTIONS

We have already seen that there are two main types of spectroscopically observable electron promotion in metal complexes, namely d–d-transitions and charge-transfer transitions. A d–d transition corresponds to the essentially angular redistribution of electrons within a d shell. In octahedral complexes, this redistribution often corresponds to the occupation of M–L antibonding e_g orbitals. An example is the $^4T_{1g} \leftarrow {}^4A_{2g}(t_{2g}{}^2 e_g{}^1 \leftarrow t_{2g}{}^3)$ transition in $[Cr(NH_3)_6]^{3+}$, The occupation of the antibonding e_g orbital results in a quantum yield close to 1 (specifically 0.6) for the photosubstitution:

$$[Cr(NH_3)_6]^{3+} + H_2O \xrightarrow{hv} [Cr(NH_3)_5(OH)_2]^{3+} + NH_3$$

This is a prompt reaction, for it occurs in less than 5ps.

Charge-transfer transitions correspond the *radial* redistribution of electron density. They involve to the promotion of electrons into more predominantly ligand orbitals if the transition is metal-to-ligand or into orbitals of greater metal character if the transition is ligand-to-metal. The former process corresponds to metal oxidation and the latter to metal reductions. These excitations commonly initiate photoredox reactions of the kind already mentioned in connection with Co(III) and Ru(II).

Although it is a very useful first approximation to associate photosubstitution and photoisomerization with d–d transitions and photoredox with charge transfer transitions, the rule is not absolute. For example, it is not uncommon for a charge transfer band to result in photosubstitution by an indirect path :

$$[Co^{111}Cl(NH_3)_5]^{2+}\ H_2O \longrightarrow [Co^{11}(NH_3)_5(OH]^{2+} + .Cl$$

$$[Co^{111}(NH_3)_5(OH_2)]^{2+} + .Cl \longrightarrow [Co^{111}(NH_3)_5(OH_2]^{3+} + Cl^-$$

In this case, the aqua complex formed after the homolytic fission of the Co–Cl bond is reoxidized by the Cl atom. The net result leaves the Co oxidized and substituted. Conversely, the long lived ^2E state of $[Cr(bipy)_3]^{3+}$ is a pure d–d state which lacks substitutional reactivity in acidic solution. Its lifetime of several microseconds allows the excess energy to enhance its redox reactions. The standard potential a good oxidizing agent, in which it undergoes reduction to $[Cr(bipy)_3]^{2+}$.

TRANSITIONS IN METAL-METAL BONDED SYSTEMS

We encountered the $\delta^* \leftarrow \delta$ transition of multiple metal-metal bonds and it should not be surprising that population of an antibonding orbital of the metal-metal system can sometimes initiate photodissociation. It is more interesting that such excited states have been shown to initiate multielectron redox photochemistry.

One of the best characterized system is the dinuclear platinum complex $[Pt_2(\mu\text{–}P_2O_5H_2)_4]^{4-}$ called informally "PtPOP". There is no metal- metal bonding in the ground state of this Pt(II) Pt (II) d^8–d^8 species. The HOMO–LUMO pattern shown in Fig. 6.6 indicates that excitation populates a bonding orbital between the metals. The lowest lying excited state has a lifetime of 9 μs. It is a powerful reducing agent, and reacts by both electron and halogen atom transfer. The most interesting oxidation products are Pt(III) – Pt(III) that contain X^- ligands (where X is halogen or pseudohalogen) at both ends end a metal-metal single bond. Irradiation in the presence of $(Bu)_3$ SnH gives a dihydrido product which can eliminate H_2.

Irradiation of the quadruply bonded dinuclear cluster $Mo_2[O_2P(OC_6H_5)_2]_4$ at 500 nm in the presence of $ClCH_2CH_2Cl$ results in production of ethene and the addition of two Cl atoms to the two Mo atoms, with a two-electron oxidation. The reaction proceeds in steps of one electron, and requires a complex with the metals blocked by sterically crowding ligands. If smaller ligands are present, the reaction that occurs instead is a photochemical oxidative addition of the organic molecule.

SPECTRA OF COMPOUNDS WITH METAL-METAL BONDS

Compounds which contain metal-metal bonds are intensely coloured. For example, $Mn_2(CO)_{10}$ is bright yellow, $Re_2Cl_8^{2-}$ is deed blue, $Co_2(CO)_8$ is purple-black, $Mo_2Cl_8^{4-}$ is cherry red, $Fe_2(CO)_9$ is gold. The MO energy level diagrams are appropriate for these compounds; the observed transitions are of the types $\sigma \rightarrow \sigma^*$, $\pi \rightarrow \pi^*$, and $\delta \rightarrow \delta^*$ and fully allowed. In the case of $Mn_2(CO)_{10}$, the $\sigma \rightarrow \sigma^*$ band peaks at 29.400cm^{-1}. One expects that this excitation should be accompanied by weakening or dissociation of the Mn–Mn bond.

Indeed, ultraviolet irradiation of $Mn_2(CO)_{10}$ in carbon tetrachloride solution leads to the intermediate formation of $Mn(CO)_5$ radicals, which abstract chlorine atoms from the solvent to form the chlorine-substituted mononuclear complex.

$$Mn_2(CO)_{10} \xrightarrow[CCl4]{hv} 2Mn(CO)_5Cl$$

Photochemical cleavage of metal-metal bonds is fairly common in metal cluster compounds. However, it appears that excitation to the relatively weakly antibonding π^* and δ^* states is ineffective for such cleavage.

LIGAND FIELD THEORY

An early empirical observation with CrIII complexes was that the following rules were obeyed :

(*i*) Consider the six ligands to lie in pairs at the ends of three mutually perpendicular axes. That axis having the weakest average ligand field strength will be the one labilized, and the total quantum yield will be about that for an O_h complex of the same average ligand field.

(*ii*) If the labilized axis contains two different ligands, then the ligand or greater field strength preferentially aquates. This may be a type of *trans* effect.

As an example, the photochemistry of $[CrCl(NH_3)_5]^{2+}$ is primarily one of ammonia aquation, labelling studies showing that the *trans* ammonia is indeed the one substituted. The photoreaction is *antithermal* ; the thermal reaction in entirely one of chloride aquation. The case of $[CrCN(NH_3)_5]^{2+}$ is interesting in that the acido group is a stronger field than ammonia; as predicted by the rules, the photoreaction is one of ammonia aquation, presumably of a *cis* ammonia. While exceptions do exist (*e.g.*, *trans*–$[CrF_2(en)^{2+}]$),[3] the photolysis rules are qualitatively consistent with the behaviour of a large number of mixed-ligand Cr^{III} complex.

The rules have been accounted for in terms of ligand field theory, taking Q°1 to be the reactive state. Figure 6.7 illustrates the energy level orderings of the one-electron d – orbitals and the electronic states for $[CrX(NH_3)_5]^{2+}$ complexes of effective C_{4v} symmetry. The lowest quartet excited state is $^4E(a)$, a component of the $^4T_{2g}$ state in O_h symmetry. Excitation of the complex to $^4E(a)$ places an electron in a strongly σ–antibonding orbital which is mainly $d_z{}^2$ in character.

2

Isolation of Zintl Anions

INTRODUCTION

The isolation of any solid derivatives of the Zintl anions has only been accomplished in the 15 years. Kummer and Diehl obtained the first example in the compound $(Na)_4(en)_7Sn_9$(en = ethylenediamine). They found the alloy composition $NaSn_{24-25}$ slowly dissolved in en (or other polyamines) on wanning and the indicated compound could be precipitated on the addition of monoglyme or THF. The reddish brown product was strongly diamagnetic and gave a single, broad ^{119}Sn Mossbauer resonance at +0.15 mm^{-1} s^{-1} vs. α-Sn with a half-width of 1.90 mms^{-1}. Additional examples of Zintl's "polyanionic salts" later reported were the copper-coloured $Na_4(en)_5Ge_9$ and the dark red $Na_3(en)_4Sb_7$ which were isolated with similar techniques and likewise identified by elemental analyses. The former slowly loses en at room temperature.

A salt of the lead anion could not be isolated. A subsequent incomplete crystal structure study of the above tin compound identified a distorted and evidently somewhat disordered Sag polyhedron therein, but then molecules could not be located well (R = 11.7%, Mo Kα, 26 $\leq$ 30°). We will return to the geometric features of this anion in a later comparison with other Sn, polyhedra that have been structurally identified.

A more general solution to the isolation of diverse examples of Zintl anions lay in the use of a more effective sequestering agent for the countercation. The all-too-ready reversion of this type of salt to an inter-metallic phase can be viewed as the result of delocalization of electrons from the Zintl polyanions-usually very good reducing agents-back onto the cations. Blocking this pathway thus, promises some stabilization.

The selection of the macrocyclic ligand I, abbreviated 2,2,2-crypt or hereafter just crypt, was guided by the reports by Dye and co-workers that this octadentate agent not only greatly enhanced the solubility of the alkali metals themselves in various solvents but also allowed the isolation of the remarkable (crypt-Na^+)Na^-. Its presence was found not only to greatly increase the rate of solution of many intermetallic phases, especially in en and tn a lesser extent in NH_3, but also to allow the ready isolation of many salts containing the Zintl anion, for example

$$2KPB_{2.5} + 2(2,2,2\text{-crypt}) \xrightarrow[50°C]{en} (2,\ 2,2\text{-crypt-}K^+)_2Pb_5^{2-}$$

The principal homoatomic anions that have been so obtained and characterized by single-crystal X-ray means together with some isoelectronic analogues. Though 2,2, 2-crypt is ideal for an ion the size of K^+, it works quite well with Na^+ as well. In this connection, it has been observed that crypt – K^+ phases tend to enclose more non-bonded en solvent whereas the sodium analogues are more apt to be solvent-free and in a higher symmetry cell. Large cations of these types often dominate the crystalline packing and sometimes allow for the disorder among the anions, or provide poor selectivity at a given site. Synthetic explorations with 2,2,1-crypt-Na^+ or 2,1,1-crypt-Li (the numbers identify the ether oxygens in each chain) have generally been less productive; these reagents generally need to be handled as solutions, and they tend to yield oils, tars, etc. rather than crystals, in part perhaps because of crypt reduction.

Comparison of the polyatomic anions that have been isolated and characterized structurally with those detected by Zintl reveals many differences. The possible identity of the isolated Bl_4^{2-} and the analysed $Bi_{5.85}^{3-}$ (= $Bi_{3.90}^{2-}$) was noted earlier. Some of Zintl's ions have been subsequently isolated by fairly conventional means, for example, the barium salt of As_7^{3-} by more direct, high-temperature reactions. Even for tellurium where the stability of polyanions might be the most borderline among the chalcogenides Te_3^{2-}, Te_4^{2-}.2MeOH, Te_5^{2-}, and the novel [$Te_4Te_{2/2}^{2-}$] have all been isolated. All of the polychalcogenide anions obtained in solution appear to be simple chain fragments and will not be considered further.

Some ions reported by Zintl and not yet isolated involve higher charges per atom than in species obtained to date, As_3^{3-} Sb_3^{3-}, Bi_3^{3-}, As_5^{3-}, and Bi_5^{3-}, for example, and their preparation in solid compounds may require either a solvent with a higher dielectric constant or a solvent or crypt-like agent that is more stable to reduction. (Some of the alloys studied reduce en to some degree with the evolution of hydrogen, more so for those with higher alkali-metal content, and the same is to be expected for certain solutes. This alloy oxidation can be useful in generating finely divided and more reactive heavy elements). Mixed element clusters are also a useful means for reducing cluster change. One of the surprises in the absence of any salt of Pb_9^{4-} even though this was the first such polyanion discovered. Dark brown or brown-green solutions are readily obtained with crypt in en from phases near the compositions $M^1Pb_{2.2}$ (M = Na, K), but only the starting alloy with the obviously very stable M^1Pb_{3-x} structure is recovered on solvent evaporation.

The use of heat or of more sodium-rich alloys provides salts of the lower charged Pb_5^{2-} which, it will be noted, has very nearly the same composition ($Pb + Pb_9^{4-} \rightarrow 2Pb_5^{2-}$). The more familiar green solution is obtained from $M_1Pb_{2.2}$ and crypt in ammonia, but again only the Pb_5^{2-} products can be isolated therefrom. No reaction occurs with the alternative $SrPb_3$. Likewise, a dark brown solution is obtained by extraction of $NaPb_{2.25}$ with en alone, but only the original alloy precipitates on solvent evaporation or on the addition of monoglyme. According to cryoscopic and conductometric measurements, Na_4Sn_9 and Na_4Ge_9 are hardly dissociated in en, even in dilute solution, perhaps accounting for some of the colour differences seen for the lead anion. The simpler Pb_5^{2-} ion, or any other polylead ion except Pb_9^{4-} for that matter, has not been detected in appreciable amounts in NMR studies on en solutions (below).

Although Na-Si alloys show no trace of solution in en or NH_3 alone, the addition of crypt to the latter gives dark green-brown solutions, and thin yellow to brown crystals can be obtained on very slow solvent evaporation at low temperatures. However, these are not stable for long at room temperature, and no structure solution (or analysis) has been attempted. Although, negative results are obtained with crypt and alkali-metal alloys or mercury or thallium, an interesting product is obtained with those of gold. In the presence of crypt, K-Aualloys dissolve in NH_3, and a white solid containing crypt and the alloy components can be recovered on evaporation of solvent at low temperatures.

This product, which decomposes above –10°C to produce metallic gold, has tentatively been suggested to be (2,2,2-crypt-K^+) Au^- containing the $5d^{10}\,6s^2$ pseudohalide anion Au~. The presence of a monoatomic gold anion in the (crypt-M^{+}) Au^- compounds (M = K-Cs) has been given strong support by a recent Mossbauer study at low temperature which reveals a single gold resonance at 7.5-8.3 mm s^{-1} relative to pure gold. (The yellow, semiconducting CsAu gives the same shift). An Au_x^- anion has also been generated and studied in liquid ammonia alone.

A collection of some of the anion structures determined by single-crystal X-ray studies of these phases namely, for Sn_9^{4-}, Sn_9^{3-}, Ge_8^{2-}, Pb_5^{2-}, Bi_4^{2-}, and Sb_7^{3-}, all of which represented new homopolyatomic anion configurations and several, the first of a kind: the Sn_9^{4-} unicapped antiprism, the Pb_5^{2-} trigonal bipyramid, and Sn_9^{3-}a paramagnetic example. Most of these will be considered later along with additional physical measurements, cluster bonding regularities and theoretical results, most particularly as the last relate to a noteworthy configurational problem, the contrasting C_{4v} symmetry observed for Sn_g^{4-} but D_{3h} for the isoelectronic Bi_9^{5+} and the unusual Sn_9^{3-}.

Prior to some general discussion, it is advantageous to also note the related heteroatomic clusters that have been obtained by similar techniques.The use of mixed post-transition elements not only provides some additional configuration examples but also allows the inclusion in clusters of electron poorer elements to the left of group 4 that are missing from homoatomic examples.

The results to date are listed in Table 1 and several are namely, $Pb_2Sb_2^{2-}$, $Tl_2Te_2^{2-}$, $TlSn_9^{3-}$, and $TlSn_9^{3-}$. Reactions between solute species and solid alloy appear to be particularly important in the formation and the stability of many of these mixed ions. In several cases, mixed species are obtained when crystals are grown in the absence of alloy, but mainly homoatomic clusters are produced after extended periods in its presence.

Table 1 : Heteroatomic Zintl Anions of Known Structure Stabilized with Crypt.

Anion	*Symmetry*	*p electrons*	*Isoelectronic alalgoues*
$Sn_2Bi_2^{2-}$	~Td	12	Sb_4,Bi_4
$Pb_2Sb_2^{2-}$	~Td	12	
$Tl_2Te_2^{2}$	C_{2v}	12	Ge_9^{2-c}
$TlSn_8^{3-}$	$C_{2v}(D_{3h})$	20	
$TlSn_9^{3-}$	$C_{4v}(D_{3h})$	22	$B_{10}H_{10}^{2-}$
$As_2Se_6^{3-}$	~C_{2h}	32	
$As_2Te_6^{3-}$	~C_{2h}	32	

Although, one's initial reaction to the prospects of forming mixed-metal species might be that heteroatomic examples would be destabilized by intrinsic differences of the component atoms and the resultant lessening of delocalization, this does not seem to be serious, at least between elements in neighbouring groups. In addition, many polyhedral examples have non-equivalent positions whereby heteroatoms presumably can be better accommodated. Even the nominal tetrahedron can be obtained with appreciably different atoms, $Sn_2Bi_2^{2-}$ and $Pb_2Sb_2^{2-}$ evidently representing the first "tetrahedral" ions stable in solution or as normal salts although nominally higher charged examples are known in the condensed solid state in most M^IM^{IV} phases. Although the examples just listed were chosen so that the atoms could be distinguished by X-ray diffraction, the anions in fact are found to be orientationally rather well disordered in the potassium salts. There is little doubt that the other obvious possibilities $Sn_2Sb_2^{2-}$ and $Pb_2Bi_2^{2-}$ also occur and will probably be disordered in such solids. Many more mixed examples can be expected as this approach provides a means to achieve ions of a given electronic configuration with a relatively low charge, for example, those isoelectronic with a variety of polychalcogenide cations.

Recent discoveries indicate that there are also a considerable variety of additional heteropolyatomic anion examples that provide a more-or-less natural bridge between examples of the sort cited in Table 4.3 where the bonding is often non-classical (multicenter) and new varieties of classically bonded and sometimes centred heteroatomic anions. Examples of the latter include solids containing $Sn_2Te_6 4^-$ (D_{2h}), $SnTe_4^{4-}$(T_d), and a variety of novel gold and mercury telluride anions such as $Hg_4Te_{12}4-$ and $KAu_9Te_7^{4-}$. There is also NMR evidence for $Te_2X_2^{2-}$ and for the centred ions HgX_2^{2-}, CdX_2^{2-} (linear), TIX_3^{3-}, and SnX_3^{2-} (D_{3h}) where X = Se and Te, all in the presence of crypt. The $HgTe_2^{2-}$ member of the last series has been isolated as the 2, 2, 2-ciypt-K^+ salt. Members of the previous groups of mixed ions have also been termed "Zintl ions" by some investigators by the criterion that they were obtained from Zintl phases. The last are polar intermetalic or salt-like phases which were also first generalized by Zintl and which will be described briefly. This group constitutes a large number of compounds that appear to blend into more conventional polyatomic ions in the limit. For the purposes of this review, Zintl ions considered in detail will be limited to (empty) cluster and related polyhedral species, often with non-classical bonding, of the character first broadly illuminated by Zintl. Even this border is not clearly defined.

The results of Mossbauer studies on crypt salts of three ions Sn_5^{2-} Sn_9^{4-}, and $Sn_2Bi_2^{2-}$ that have been structurally characterized as the crypt salts are consistent with the known structures. All isomer shifts are in a range of δ = 2.25–3.00 mm s^{-1} vs. $CaSnO_3$. The asymmetric line obtained at 77 K for the Sn_5^{2-} salt resolves into two components at the extremes of this range. Although these have an intensity ratio of 3.8:1 rather than the expected 1.5:1 for the D311 ion, the temperature dependence of this ratio is positive and sizable, a change which the authors attribute mainly to the more weakly bound apex atom that gives rise to the more weakly bound apex atom that gives rise to the smaller component. Very consistent intensity ratios of the order of 1.5–1.8 can be obtained by graphical extrapolation of the data to 0 K. The Sn_9^{4-}spectrum obtained at 77 K resolves best into a 2.5:1:2.5 population triplet, which was considered to be in good agreement with the expected 4:1:4 relationship, while $Sn_2Bi_2^{2-}$ salt gives the expected single transition and the only example of quadrupolar splitting (1.07 mms^{-1}). Comparison of the second of these with the reported Mossbauer results for a $Na_4(en)_7Sn_9$, is not very close, probably because of a somewhat different tin polyhedra, evident Sn-Na interactions and some unavoidable oxidation of the sample in the earlier study.

STABILITY PRINCIPLES AND COMPARISONS

Although it is clear that synthetic problems associated with the isolation of many Zintl anions are substantially lessened through the use of the crypt complex for the cation, this may not always be necessary for stability. Isolation of Si_9^{4-} as well as the probable Sb_7^{3-} and Ge_9^{4-} ions in salts where the sodium cations are more or less sequestered by en has already been noted. Teller et al. have recently obtained both $(Me_4N)_4Sn_9$(N:Sn=4:9.4)asared-brown powder from the rapid reaction of the appropriate potassium alloy with a Me_4N^+ solution in DMF and a well-crystallized $[K\text{–}(HMPA)_2]_4Sn_9$ by a like reaction using HMPA in en solvent. However, both phases decompose at room temperature, the former to Me_3N and Me^xSn. The use of crypt to gain $HgTe_2^{2-}$ as well as Te_3^{2-} probably represents only a convenient route to a large cation and is not a necessity. The same may be true with other anions such as $As_2Se_6^{2-}$ and $Tl_2Te_2^{2-}$.

The postulated effect of cryptation in blocking electron delocalization back into the cation clearly is altered when the solid decomposition products are more salt-like than metallic, with NaSb and Na_3Sb, for example, and more general questions of phase stability now apply. Other pertinent factors include the well known advantages of using a large cation with polyatomic anions of limited stability, the improvement in solubility of higher charged ions in solvents with modest dielectric constants such as en that is afforded by larger cations, and simply appreciable changes in solubility patterns. Either the last is fairly significant or the crypt complexation drastically alters the stability of ion pairs judging from the isolation of crypt salts of species such as Sn_5^{2-}, Ge_9^{2-} and Sb_4^{2-} that have not been evidenced in solution or in solids obtained in the absence of crypt. These advantages for whatever the reasons have led to the apt description of crypt as a "synthetic laxative".

It is noteworthy that the isolated Zintl ions have several isoelectronic analogues among the polyatomic cations investigated earlier, viz., Bi_5^{3+} relative to Pb_5^{2-} and Sn_5^{2-}, Te_4^{2+} compared with Sb_4^{2-} and Bi_4^{2-}, and Bi_9^{5+} vs. Sn_9^{4-} and Ge_9^{4-}, although distinct differences in configuration between the 22-electron cations and anions in the last group will require subsequent attention.

Notwithstanding, the considerable contrast in the chemical systems used in isolating the polycation vs. The polyanion members is worth noting. Both in effect depend on the nature of the counterion. The tendency for disproportionation into element and normal-valent ions shown by the cations, for example

$$Bi_5^{3+}(X^-)_3 \rightleftharpoons BiX_3 + 4Bi$$

can be greatly reduced by the use of a large anion, preferably a nonoxidizing one such as AlC_4^-, $HfCl_6^{2-}$, etc. Reactions are run either in the molten state at elevated temperatures or in a solvent of very low basicity. This effect can be understood in terms of either the reduced difference in lattice energies of the salts with the two oxidation states or a reduction in the degree of solvation of the more acidic cation in the higher oxidation state. These effects have been viewed as both "acid stabilization" and "anticoordination chemistry". On the other hand, the disproportionation of the Zintl anions by a reaction like

$$(Na^+)_2Bl_4^{2-} \rightleftharpoons 2NaBi + 2Bi$$

is driven by the stability of NaBi and the fixed heat of condensation of the elemental Bi. In this case the idea is to sequester the countercation sufficiently that the decomposition is unfavourable, the metallic NaBi product, in the above being replaced by an unknown and probably unstable version $NaL_n^+ Bi^-$ Although the initial interpretation of the effect of crypt was one of preventing electron transfer or delocalization back into the countercation, clearly this role is different when the product is not metallic, for example with NaSb, and we are really back to the effect of an alteration in both cation size and bonding ability on general phase stability.

SOLUTION STUDIES

Nearly all modem information on the solution chemistry of the Zintl ions has come from the NMR studies originated by the late R. W. Rudolph. The majority of these were carried out in en without the addition of crypt; although the solution processes are then slower, more concentrated systems are ultimately achieved in many systems. A convincing proof of the predicted fluxionality of Sn_9^{4-} was supplied by the observation by pulsed Fourier transform NMR methods of a single ^{119}Sn resonance 1230 ppm upfield from Me_4Sn, this resonance being split by ^{119}Sn– ^{117}Sn coupling into the central five lines of the multiplet intensity distribution predicted for a nine-atom cluster.

Thus, the three non-equivalent atoms expected on the basis of what is presumed to be the C_{4v} ground-state ion found in the solid are averaged in an intramolecular manner on the NMR time scale, possibly via $C_{4v} \rightleftharpoons D_{3h}$ interconversion pathway (see below). The tin spectrum is unchanged at –40°C in liquid NH_3. The $Sn_xPb_{9-x}^{4-}$ and $Sn_xGe_{9-x}^{4-}$ familites of isoelectronic ions were also observed to form with appropriate ternary alloys. Portions of the group of spectra observed for the members of the $Sn_xPb_{9-x}^{4-}$ family by ^{119}Sn NMR and by ^{207}b NMR, respectively. These and the observed Sn-Pb coupling establish that the complete series of nine atom clusters is present. Since the shift appears to be quite regular the series was taken to be that of the 4- ions throughout.

The chemical shift for tin in this series was found to depend linearly on the electronegativity-corrected charge per tin, thus, giving an approximate means of predicting the charge per tin on unknown ions (provided paramagnetic and diamagnetic contributions to the shift are relatively invariant). Incidentally, the homogeneous equilibration of Sn_9^4 and Pb_9^4 ions that had been previously prepared separately was observed to take weeks at 76°C but to be much more facile in the presence of a ternary tin-lead alloy. The same family of tin-lead clusters may be generated electro-chemically and without the serious tin passivation observed in liquid ammonia. All members of the analogous $Sn_2Ge_9^{-4}$ series were also found but without observable Sn-Ge coupling (I = 9/2 for Ge). The variation in chemical shift with x in this series is distinctly smaller and changes sign.

Table 2 : Zintl Ions Deduced from Solution Studies.

Ion	*Study*	*Result*
$Sn_xPb_9–x^{4-}$	^{119}Sn, ^{207}PbNMR	all species $0 \leq x \leq 9$; fluxional
Sa_xGe_{9-x}	Sn NMR	all species $0 \leq x \leq 9$
Sn_4^{2-}	^{119}Sn NMR	single resonance for four atoms, J = 1217 Hz
	Na:Sn solution analysis by NMR effective potential calculations	2:1 fluxional via D_{2d}"tetrahedra"
Sn_8Sb^{3-}	^{119}Sn NMR	fluxional $Sn_8Sb_x^{n-}$, x assumed, n based on shift and Sn_9^{4-}
$TISn_8^{5-}$	^{119}Sn, ^{205}TI NMR in en and NH_3 solution analysis	fluxional $TISn_8$ composition; charge based on Sn_9^{4-} $Na_5(en)_{1.25}$ $TISn_8$
	^{119}SnMossbauer on solid	broad line, ambiguous
$Sn_2Bi_2^{2-}$	^{119}Sn NMR, analysis of LiSnBi solution	$Sa_2Bi_2^{2-}$ (δ = –1665, J = 1606 Hz)
As_4^{2-}	EXAFS on solution of K_3As_5	$As_4^{2-}(D_{4h})$ + $As_6^{4-}(D_{6h})$, 50:50

Changes in the chemical shift of a given species as a function of cation in these solutions clearly manifest the presence of ion pairing. For example, the ^{119}Sn shifts (vs. Me_4Sn) for Sn_9^{4-} in en vary rather smoothly from –1115 ppm for Cs^+ to –1241 ppm with Li^+ and –1253 ppm with 2,2,2-crypt –Na^+ the larger crypt or the well-solvated lithium being similar in effect. The comparable range for the ^{207}Pb shift in Pb_9^{4-} with alkali metal is –891 to –1224 ppm.

Lists the above together with other ions for which some appreciable evidence has been obtained from solution studies, mainly NMR. One of the more remarkable is the result of equilibration of alloys in the $NaSn_{1-2}$ range with en. The dilute, yellow-red solution contains a high yield of a species with a single ^{119}Sn resonance (δ – 1895) split by ^{117}Sn coupling (J = 1224 Hz) in a manner clearly appropriate to a four-atom cluster. These data suggest either a tetrahedron or a lower symmetry fluxional species, the former being the more reasonable for such a large J value.

A rough NMR analysis using internal ^{23}Na and ^{119}Sn standards together with the shift-per-charge on tin are consistent with an 18-electron Sn_2^{2-} species. Of course, a normal tetrahedral unit would be expected to be paramagnetic and Jahn-Teller unstable.

On the other hand, fluxionality of a diamagnetic species about a tetrahedral average is supported by effective potential calculations for Sn_4^{2-}, dynamic second- order Jahn-Teller distortion producing a fluxional diamagnetic system that circulates via pseudorotation among the six D_{2d} configurations, *i.e.*, elongated and compressed tetrahedra. The calculated barrier is 40 kJ mol^{-1}. Observation of the limiting spectrum of the rigid ion at lower temperatures would certainly be valuable. Some evidence for a similar ion for germanium has been obtained from an X-ray study of what was apparently $(2,2,2,\text{-Crypt-Na}^+)_2Ge_4^{2-}$. Although, a tetrahedral-like anion seemed evident (d-(Ge-Ge) ⋦ 2.78 A, B ≈ 11 A), the anion was partially disordered about an effective inversion centre, and the structural solution was not really completely satisfactory (R = 17.0%). (The preliminary report of a crystalline salt of Sn_4^{2-} was in error.)

A spectrum attributed to a fluxional $TlSn_8^{5-}$ was obtained from M_2Sn_2Tl alloys (M = K, Na) in both en and NH_3, the composition being uniquely established via Sn-Sn and Sn-Tl coupling. The 5-charge was based on a total solute analysis $Na_5Sn_{8.4}Tl$. 1.25en for the residue isolated by evaporation of en. The charge assignment is of course at variance with the one isolated example, $TlSn_8^{3-}$; the latter also involves heteroatom substitution at a lower order vertex contrary to prediction for the electron-poorer thallium in the 5-ion. The low en content of this solid obtained presumably allows sizable Na-Sn interactions, and these lead to the rather broad and uninformative ^{119}Sn Mossbauer spectrum obtained for this material.

Equilibration of a variety of Na-Sn-Sb compositions with en gives rise to a new species that contains eight tin atoms and is also fluxional according to tin NMR (J = 81 Hz) (an earlier analysis assigned this to nine atoms). No Sn-Sb coupling was seen, presumably because all antimony isotopes are quadrupolar. Although the antimony content could not be determined, a reasonable Sn_8Sb^{3-} ion isoelectronic with Sa_9^{4-} was suggested. However, the ^{119}Sn shifts found for Sn_8Tl^{5-} and Sn_8Sb^{3-}, 61 and 108 ppm downfield from (less negative than) Sn_9^{4-} for the sodium salts, would certainly be more consistent with expected electronegativity equalization effects and the shift per charge trends found for the Sn_x

$Pb_{9-x}{}^{4-}$ series if the former were instead Sn_8Tl^{3-}, the same as known structurally. A plausible fluxionality in this event is not clear.

Other ions have been postulated on the basis of similar or less NMR evidence, Sn_8Bi^{3-} (mixed with $Sn_2Bi_2{}^{2-}$), for example. There are also some fascinating but incompletely characterized species obtained by reactions of $Sn_9{}^{4-}$, $Pb_9{}^{4-}$, or $TlSn_8{}^{5-}$ with $Pt(PPh_3)_4$ or $Pd(PPh_3)_4$ in en. The first pair of reactants give the best characterized product $(PPh_3)_2PtSn_9{}^{4-}$ the charge being assumed, while $Pb_9{}^{4-}$ forms a seemingly analogous species. The reaction of $TlSn_8{}^{5-}$ was though to give $(PPH_3)_2MTlSn_8{}^{5-}M$ = Pt, Pd, or similar species; although the 2,5 and 8 numbers in this formation were in effect assumed, the ^{119}Sn chemical shifts are very similar to those of the dinuclear products above. The thallium-containing species subsequently exchanges with tin, evidently at the surface of the alloy, to form the previously known $(PPh_3)MSn_9{}^{4-}$ ions. The species are evidently not very stable at room temperature. Additional supporting evidence is needed but the prospects of an interesting chemistry are manifest.

The one EXAFS study indicating the existence of $As_4{}^{2-}$ analogous to its antimony and bismuth brethren that have been characterized in the solid state will hopefully be followed by other studies of this character, for example, on the polybismuth solutions which, based on colour changes, should contain more than just the isolated $Bi_4{}^{2-}$ ion ($Bi_6{}^2 = Te_6{}^2$ + ?). This means of analysis will naturally be facilitated if auxiliary information about solution composition and the number of species present is available.

Some novel applications of Zintl ion solutions should be noted. Oxidizing polymers such as a polyimide can be topochemically metallized by reaction with such solutions, the alkali-metal cations intercalating as the anion reduces the polymer to anion centers while thin films of the oxidized (neutral) metal from the anion are simultaneously deposited of the surface. Solutions of K_4Sn_9 in en reduce graphite, $HfTe_2$ and the like, the potassium being intercalated while tin metal deposits only on edges of the layered substrate.

CONFIGURATION, BONDING AND ELECTRONIC CONSIDERATIONS

The so-called "naked" clusters formed by the postransition elements are in most respects not something apart in configuration and bonding. As is usually the case, certain bonding arrangements are favoured and occur in diverse compounds, and a close similarity in skeletal electron count and bonding between these clusters and the boranes, $B_nH_n^{2-}$ particularly, has been recognized for some tune. Thus, there is little that is very unusual about the configurations found for most of the Zintl ions, particularly for the more electron poor and clustered group of these, although some provide new examples of predicted configurations.

The polyhedral shape-electron count relationships generally fall into the well-known categories described by Wade's rules, namely, the deltahedral or closo type with n atoms bonded by In + 2 electrons (Ge_9^{2-}, Sn_9Tl^{3-}, Pb_5^{2-}), the nido species with In + 4 skeletal electrons where one vertex has been removed from the closo configurations for n +1 atoms (Sn_9^{4-}), and the arachno members with still one fewer vertex and 2n + 6 electrons (Sb_4^{2-}, Bi_3^{2+}). The long standing exception to these regularities has been the closo but 2n + 4 ion Bi_9^{5+} (tricapped trigonal prism), the explanation of which will be considered shortly. (The tetrahedral cluster limit also requires a little fancy footwork-in the above terms the well-behaved tetrahedral species with 2n + 4 electrons exemplified here by $Sn_2Bi_2^{2-}$ is viewed as the nido product of a trigonal bipyramid; the proper 2n + 2 closo tetrahedron would be paramagnetic and Jahn-Teller unstable, but, interestingly, such a diamagnetic species Sn_4^{2-} has been proposed to be fluxional about the ideal configuration.)

The distribution of species in terms of p electron count per cluster atom for the structurally characterized Zintl anions, some cations, and the polyboranes are function of period. The 2n +1 lower limit is clearly seen at ~ 2.25 skeletal electrons per atom (2 + 2/n more precisely). Incorporation of very many group II or III elements into these ions presumably must be accompanied by enough of an electron richer element to achieve this minimum. (The obviously special type of metal-metal bonding shown by mercury is a notable exception). At the other extreme, valence electrons in the mainly cationic clusters with more than about three electrons per cluster (2n +6 is closer) can be partitioned into ~ 2.2-3 skeletal bonding electrons with the remainder appearing as the nonbonding lone pairs that are so important in determining the shape of the molecule or ion. The latter occur in an even higher proportion in the unlisted polychalcogenide anions with more than four electrons per atom.

It is worth noting that in spite of the relatively low coordination numbers and electron counts in these Zintl anions, they do not act electron deficient in the sense of being strongly electrophilic or acidic, in fact, the only evident interaction of en with the anions in all the solids studied is the apparent hydrogen bonding of en to Te_3^{2-}. In contrast, Te_4^{2+} and Se_4^{2+} clearly interact with such poor bases as $SbFe_6$, AlC_4^-, and $Al_2Cl_7^-$ in their solid compounds. Orbital-wise, the relatively negative open face of Sn_9^{4-} is quite analogous to the products obtained either by vertex elimination from deltahedral $B_nH_n^{2-}$ or the related carboranes $C_2B_{n-2}H_n$ or by their reductive polyhedral expansion except that the open face is closed and stabilized by cobalt, iron, etc. atoms in the latter in carbollide-type complexes. The tin cluster is obviously stable without such a feature, but the possibility of analogous reaction of the open face is still intriguing and unexplored.

Only a few of the clusters previously described are amenable to classical two-centre bonding descriptions. Qualitative multicenter schemes can be applied to Pb_5^{2-} and Bi_9^{5+} viz., eight three-center and three four-center bonds in the latter. But for the most part, only molecular orbital approaches have provided general and energetically useful descriptions. Some of the distinctive close-shell or saturated bonding characteristics of these Zintl clusters and the analogous cations can be understood in these terms as originating with the relatively small, atomic orbitals basis set involved. This is in most cases effectively orbital only with MO's derived from valances being effectively core states, a factor that was first recognized in the analogous polybismuth cations. Obviously, such a circumstance makes approximate molecular orbital calculations both easier and more creditable although the intrusion of important relativistic effects with the heavier elements may "even the score".

In fact, an apparently meaningful factor that is not evident in conventional extended Huckel calculations is the significantly diminished binding energy per atom calculated for a variety of anionic clusters of sixth period elements, lead, for example, that is found on inclusion of spin-orbit coupling. This predicts a binding energy for polylead ions only about 60% of that calculated for the analogous tin species, with those for bismuth and germanium intermediate. These binding energies are naturally only part of the story in a cyclic description of decomposition reactions, in fact, another term, the heats of atomization of the individual elements, somewhat parallels the calculated cluster binding energies, a rather sensible result in itself.

By and large, the published collection of calculated molecular orbital energy results agree fairly well with observation as far as ground-state configurations. But a significant problem in this sort of prediction is the choice of dimensions of unknown configurations that perhaps are only slightly higher in energy. This will be evident in the following consideration of some characteristics of nine-atom species.

PROBLEMS WITH NINE-ATOM POLYHEDRA

Two features of the 22-electron cluster examples have stirred considerable interest; their fluxionality and the contrasting configurations of C_{4v} for Sn_9^{4-} and D_{3h} for the rule-breaking Bi_9^{5+}. On the other hand, the 20-electron polyhedra typified by Ge_9^{2-} and $TISn_8^{3-}$ are properly closer (D_{3h}) or approximately so, and low-energy alternatives are not known to exist. Some helpful clarifications regarding important factors in the configurational dilemma for 22-electron, nine-atom polyhedra are provided by a combination of the recent isolation of the intermediate Sn_9^{3-}, which is also D_{3h}, dimensional comparisons among the D_{3h}, polyhedra, and some EHMO calculational results.

The configurational assignment of many nine-atom polyhedra must be decided with some care since the two most common limits, the unicapped archimedean antriprism (C_{4v}) and the tricapped trigonal prism (D_{3h}), are not very far apart in either distances or angles; in fact, they can be easily confused or misjudged if they are classified solely by inspection of drawings made from selected perspectives. The bases for their differentiation can be found which also illustrates the probable pathway for a $C_{4v} \rightleftharpoons D_{3h}$ interconversion between these two though a C_{2v} intermediate and along a suitable extended vibration of each species. The left view is along the four fold axis of the C_{4v} model with the square base toward the reader.

Generation of a D_{3h} polyhedron with the three fold axis vertical can be accompanied by the indicated motions, compression of the front square along the edge 7-8 with (usually) an elongation of the 1-3 and 4-6 edges of the capped square to generate a trigonal prism. Obviously, an equivalent pathway is a generation of an alternate trigonal prism 3-6,2-5 and 1-4 with a horizontal threefold axis, which process when cycled completes the scrambling of the atoms. The more discriminating distances are the diagonals in each of the two squares in the C_{4v} limit, the three edges that describe the height of the trigonal prism, and the separations of the capping atoms 2,5 and 9 in the latter. More definitive, however, are the dihedral angles, particularly when clusters of different elements are involved. The more useful angles are those defining the planarity of the square base in C_{4v} (2-7-8 vs. 5-7-8, $\delta = 0°$, is one set) and one of the two pair between opposed faces around the waist, *e.g.*, 3-6-8 and 1

-4-7, that become the basal members of the trigonal prism ($\delta = 180°$). An important parameter for the D_{3h}, polyhedron is also the height-to-basal edge ratio (h/e) of the prism, *i.e.*, d(l–3)/d(l–4). The dihedral angles for the nine-atom Zintl ions and Bi_9^{5+} are sorted the most informative values described above are listed in the first two collines.

The good approximation of Sn_9^{3-} as D_{3h} is clear from the first column (and the prism heights therein differ by only 0.045 (3) A) while its difference from Bi_9^{5+} in the narrow group of adjoining vicinal angles depends on hie for the prism; $TISn_8^{3-}$ is certainly similar. At the other extreme, Sn_9^{4-}, (and KSn_9^{3-} later) is close to C_{4v} and the best trigonal prism heights differ by more than 0.9 A. The two germanium polyhedra are distorted from the end members Ge_9^{4-} is clearly derived from C_{4v} (the prism heights differ by 0.6 A) while Ge_9^{2-} though seemingly derived form D_{3h} is (inexplicably) distorted 25-30% toward C_{4v}, one prism height now being 0.33 A longer than the other two. (This does not appear to be along a suitable fluxional pathway for 20-electron D_{3h} species). There is an also older structure of B_9^{5+}, the original $Bi_{12}Cl_{14}$, in which the cation is ~ 25% of the way along the same path. Crystal packing was originally invoked to "explain" this, but it is now more plausible as this distortion is probably along a low barrier or "soft" pathway;

In this connection, it should be mentioned that the tin polyhedron resolved in $Na_5(en_9)_7Sn$, is distorted in virtually the opposite direction, with two edges of what would be the best trigonal prism 0.6 A longer than the third. But in this case strong interaction of tin atoms on each of the long edges with Na^+ seems responsible, as will be described in the next section.

Calculations on the stability of different configurations for 22-electron species relevant to possible fluxionality have been carried out several times and by varying means. Two additional problems with such deductions are probably as important as the method. First, a sizable dipole for the C_{4v} alternative disappears on conversion to the D_{3h} model a factor which should bias the results in favour of the former in the usual polar molecular solvents. In contrast, Bi_9^{5+} can be observed only under acidic conditions with nonplanar anions or, in principle, also in nonbasic solvents. Second is the uncertainty regarding the proportions of unknown polyhedra, particularly the D_{3h}, version of the nine-atom Zintl anions. Scaling with simple proportions is not very realistic.

It is now known that the principal difference among various members is in the relative height-to-edge ratio (h/e) of the trigonal prism, that in

Bi_9^{5+} being 10-15% greater than that in Ge_9^{2-}, $TISn_8^{3-}$, and $B_4H_9^{2-}$, with Su_9^{3-} intermediate. This actually involves opposite changes in both the height and the basal edge, appropriate to the orbital occupied (below), and these differences can have a significant effect on the calculational results.

Some CNDO and EHMO calculations using C_{4v} and D_{3h} models that include all of these dimensional subtleties provide the energies given in Table 3 (Those for observed configurations are given in boldface type). The energies calculated for the observed configurations of these ions differs from those of the ideal limit only for Ge_9^{2-} (–0.04 eV). The CNDO predictions of configuration for M_9^{4-} species are seen to favour the D_{3h} limit slightly, contrary to the known structures, and in fact the original fluxoinality predictions by this method certainly benefited from some luck in the choice of dimensional parameters. The EHMO results seem more credible in this respect and the importance of the prismatic dimensions in D_{3h} is evident (hie is given in parentheses). The $C_{4v} \rightleftharpoons D_{3h}$ interconversion clearly involves the smallest energy difference (and virtually no barrier) for the 22-electron species, as it observed by NMR for Ge_9^{4-}, and Pb_9^{4-} and implied by these data for Bi_9^{5+}.

Calculations for the unknown species Sn_9^{2-} (D_{3h}) give some evidence for its limited stability. The orbital energy sum in this instance is greatest for a 1.01 h/e ratio, decreasing by 0.7 eV at h/e = 1.15. The a_2'' LUMO for Sa_9^{2-} is rather isolated, lying within what would otherwise be a gap of 6.2 eV and only 1.6 eV above the e' HOMO. (The same is true for the known B_9H_9) This a_2'' orbital becomes the HOMO is Sn_9^{3-} and Sn_9^{4-} (D_{3h}), and in the process its energy approaches that of the other occupied MO's, A C_{4v} configuration for Sn_9^{3-} is out of the question as it would require an e^3 HOMO; the total energy also appears to be significantly less.

According to orbital energy date, the elongation of the Bi_9^{5+} ion along the threefold axis (with a simultaneous reduction of the length of the basal edge of the prism, below) clearly represents an alternative to the "correct" C_{4v} structure and experimentally is found at least in media where the dipole moment of the latter would be of no benefit. Some of the driving force for the formation of the 22-electron Bi_9^{5+} derives from the reduction of the relatively high 7+charge of the normal closo ion, a process that is disfavored for Sn_9^{4-}. That this elongation is a reasonable and feasible distortion with 22 electrons was evidently first stated in graph theory terms by King. It was later observed that the a_2'' HOMO is σ-antibonding with respect to the height of the trigonal prism, π-bonding as far as the basal edges there of, and weakly bonding of the waist or capping atoms. The first two effects can be easily seen in the character of the symmetry orbitals in 2 that, with different coefficients, go to make up this MO.

Table 3 : Energies (eV) from Molecular Orbital Calculations on C_{4v} and D_{3h} Configuration for Known Nine-Atom Polydedra.

	C_{4v}	D_{3h}, *h/e*
	CNDO-SCF Atomization Energies	
Ge_9^{2-}	54.88	55.99(1.10)
		56.62 (1.06)
Ge_9^{4-}	44.54	44.49 (1.10)
		45.85(1.15)
Sn_9^{4-}	28.61	29.10(1.15)
		28.94(1.01)
Extended Huckel (iterative)		Orbital Energies
S_9^{4-}	–311.8	–310.8(1.15)
		–309.9(1.01)
Sn_9^{3-}	–336.9	–338.5(1.08)
		–338.2(1.01)
Bi_9^{5+}	–740.0	–7410.1 (1.15)
		–739.2(1.08)

Accordingly, the principal dimensional changes that occur on occupation of this a_2" MO are not just an increase in the prism height but also a decrease in the basal edge and perhaps a tighter bonding of the capping atom. The second is clearly seen with $TISn_8^{3-}$ (face capping 71) relative to Sn_9^{3-} where the basal edge decreases from 3.12 to 3.06 A while the capping distance (omitting Tl and two partially is ordered Sn atoms in the former ion) shows only a marginal decrease from 2.96 to 2.94Å.

The demonstrated effect of the addition of two electrons to the $2a_2$" level of the nine-atom, D_{3h} polyhedron is shown in Fig. 4.10 as the observed height-to-edge ratio (h/e) as a function of the skeletal electron count. As is seen, Sn_9^{3-}, ($2a_2$"), is beautifully intermediate between $TISn_9^{3-}$ and B_4^{5+} where $2a_2$" is empty and doubly occupied, respectively, (one long prism edge each in $TISn_9^{3-}$ and Ge_9^{2-} was ,omitted in the calculation of hie, this divergence in the former case doubtlessly arising from coupling of atoms in the refinement; otherwise the hie values would be 1.03 and 1.10, respectively).

CATION-CLUSTER INTERACTIONS

The effects of ion pairing for Zintl ions in en are implicit in the NMR data, but significant interactions of this sort in most of the solid phases studied are precluded by the presence of cryptated cations. A few contrary examples do suggest that some novel structural effects remain to be discovered. Of course, these examples really just represent bridges to the type of interactions found in the solid state in Zintl phases.

Salts of two relatively stable anion obtained from en solution are known to show intimate ion association. In Rb_3As_7.3en the rubidium atoms bridge between the two-coordinate atoms of the nominal As_7^{3-} ion (C_{3v}), these being coordinated by en only on the outside of the Rb3As7 complex. A similar arrangement is found in Na_4P_{14}. 6en. Such complexes are thought to be important in the fluxionality of P_7^{3-} (by NMR) in THF and en.

A remarkable example of not-so-naked tin cluster was evidently the result of a reaction between KHgSn and an accidental deficiency of crypt in en, which yielded (2,2,2-crypt-K^+)$_3$(KSn_9^{3-}). The KSn_9^{3-} unit in this consists of an infinite chain of alternating, nominal K^+ and Sn_9^{4-} units. The non-equivalent potassium atoms lie at centers of symmetry and are quite free of coordinated en, being surrounded by the crypt cations and six tin atoms in two clusters, four at 3.55-3.76 A and two more at 4.11 or 4.22 A in a distorted antiprism. The average K-Sn distance in the first group is close to the minimum value observed about six or eight-coordinate potassium in K_4Sn_4.

Distances in the tin cluster show remarkably little evidence of the neighbouring potassium relative to those in the isolated anion in (2,2,2-crypt-Na^+)$_4$(Sn_9^{4-}). The Sn-Sn distances in KSn_9^{3-} in detail average slightly longer (0.016 A) when both atoms have potassium neighbours than for other distances of the same kind. The cluster is also slightly distorted angularly, the two nominal squares in the C_{4v} polyhedron being slightly elongated toward the potassium atoms along the diagonals; format are still close, 156°, 160° and 1°. It is certainly not clear that this arrangement represents more than an electrostatic interaction. A quite different result of the interaction of a cation with Sn_9^{4-} appears in the lower symmetry polyhedron present in $Na_4(en)_7Sn_9$. The view approximately a vertical trigonal prism that is clearly distorted away from the fluxional pathway discussed before with two edges elongated by over half an Angstrom from a slightly short value for the third (h/e as 1.06). The source of the distortion is evidently two sodium atoms (open ellipsoids) that are three rather than four coordinate to nitrogen in en and bridge the long edges of the prism. This can also be viewed as the result of capping two of the folded rhomboidal side faces of the tricapped trigonal prism, distorting both of these toward square planar although the diagonals of these still differ by ~ 0.6 A. The (rather large) thermal ellipsoids do not suggest that the resolved figure is actually the result of superposition of two capped antiprismatic polyhedra.

The observed Na-Sn distances, 3.49-3.59Å((σ = 0.02Å, are relatively about 0.2Å longer than those found in KSn_9^{3-}, when the difference in cation radii is taken into account. The magnitude of the effect with sodium suggests the barrier to this distortion of the cluster is much smaller than that in KSn_9^{3-}.

ZINTI PHASES-NEW OPPORTUNITIES

Obviously, many Zintl ion species remain to be detected and characterized both in solution and as solid salts. But the next frontier in this area may be the variety of unusual clusters or aggregates of the main-group elements that are found in the so-called Zintl phases, particular those that cannot be readily interpreted by classical valence rules.

The term "Zintl phase "was first applied by Laves to the binary compounds formed between the alkali or alkaline-earth elements and the main-group elements from group 4 on-that, is, to the right of the "Zintl boundary". (Use of the term post-transition rather than main group is also appropriate save for a few pertinent compounds of aluminium and silicon). These combinations not only yield some ZintI anions in solution but also produce many rather polar or salt-like phases.

The most reduced member for each is usually a classical valence compound in which the more noble member achieves a filled octet and an 8-N oxidation state in salt-like structure, for example, Na_3As, Mg_2Sn, and Mg_3Sb_2 which occur in LaF_2, CaF_2 and La_2L_3 antitype structures, respectively. An important intermetallic structure type discovered by ZintI was that of NaTI where the thallium sublattice occurs in the diamond structure (and the sodium likewise). This obviously can be interpreted as an isoelectronic Tl^- array in the limit of complete charge transfer. Subsequent applications of the term "Zintl phase" have generally been based on the structural characteristics of such polar phases. An expanded definition by Klemm and Busmann encompassed many more phases by including those in which the anion lattices have structures and electron counts of neighbouring elements. Examples of the latter are CaSi; (arsenic-like layers of Si), KGe, $BaSi_2$, and Na_2Tl (tetrahedra iso-electronic with P_4 and As_4), $CaGa_2$ and Cain; (graphite and distorted diamond lattices, respectively), $SrSi_2$ (a three-dimensional lattice of three-coordinate Si), and NaSb and CaSi (helical and zig-zag planar chains, respectively, analogous to gray selenium). This way of looking at electronic requirements is generally familiar to chemists, based as it is on conventional octet ideas and can be generalized to cover many filled band examples of AB semiconductors with diverse A-A, B-B, or no homotatomic bonding. Naturally, the extreme charge asymmetries implied by the oxidation state assignments do not represent real charge distributions.

A more recent and broader definition includes any heteropolar phases in which classical two-center valence rules satisfactorily describe the bonding of the post -transition elements in the "anions". It is clear that this encompasses a large and diverse group including not only such members as CaP (Na_2O_2-type), Ba_5Pb_3 (equal numbers of isolated atoms and dimers), and Sr_3As_4 (As_4^{6-} chains) but also analogues of sulfate, phosphates, silicates such as $Ba_4(GeAs_4)$, and their condensation products, some of which naturally represent new types of connectivities. The conventional bonding in many of the last category has led this author to resist a recent tendency to broaden the definition of Zintl ions to include anything obtained in or isolated from a solution of a Zintl phase.

As first noted by Zintl, substantially all of the classical Zintl phases exhibit properties reflecting the heteropolar bonding and their dissimilarity from typical alloys; brittleness, colour, relatively fixed stoichiometry, and substantial heats of formation and negative volume changes relative to the elements. All are probably semiconductors (or semimetals) although this has not been ascertained in most cases. However, it is clearly incorrect to consider the heteroatomic (cation-anion) interactions in these as ionic any more than CuSe or PbI_2 is "ionic". The use of the convenient cation and anion terms should not imply this either. An important covalency between these "ions" seem to be particularly evident in some of the extremes found within Zintl phases containing lithium.

Probably more interesting and more related to non -classical Zintl ions are the polar phases in which aggregates formed by the more electronegative elements do not readily succumb to straight forward two center bonding interpretations with conventional oxidation state assignments. In principle these may result when excess electrons make the phase metallic ($A1B_2$), when a significant participation of the electropositive element in the anion bonding scheme generates less intelligible bonding interactions, or when unprecedented clusters are found.

Two classes of compounds that show persistent and unusual aggregation of the post-transition elements even in binary phases involve either lithium or gallium or both. In the former case one finds, for instance, dimers of the heavy elements in Li_5S_2 and Li_9Ge_4,, dimers and isolated atoms in Li_8Pb_3, Li_7Sn_2, $Li_{13}Si_4$, and Li_2Sb, bent trimers in Li_7Sn_3, zig-zag chains in Li_9Al_4 and Li_2Ga, pentagons plus isolated atoms in Li_{11} Ge_6 and puckered layers in Li_3 Al_2 and LiSn. Such inexplicable bonding arrangements between the heavy elements boggle the mind.

As an example of the progress that can be made, some understanding is now available relative to bonding in the remarkable $Li_{12}Si_7$ which contains planar Si, rings and trigonal-planar Si_4 stars. The compound is semiconducting and diamagnetic, indicating a filled band with an unconventional bonding behaviour. A theoretical study shows that covalent Li-Si bonding is important, especially with the Si_4 star where sets of lithium distributed around it effectively lower the energies of the σ^* and π^* levels to allow their occupation. Similar studies of Li_8 Mg Si_6 where there are five membered rings and isolated silicon atoms show that a more normal assignment is reasonable (if a cyclopentadienide-like Si_5^{6-} limit is normal).

The variety of structures that may be provided by nature when one goes exploring is well illustrated by the remarkable group of polyhedral gallium clusters found in gallium rich binary phases formed with the alkali metals. Composition like K_3Ga_{13}, $RbGa_3$, and Na_7Ga_{13} exhibit complex structures containing large, interconnected, and usually empty gallium polyhedra; icosahedra, dodecahedra octadecahedra, etc. reminiscent of boron chemistry. Electron transfer from the alkali metal to the gallium network seems evident. The majority of the arrangements may be interpreted as Zintl phases fairly well with the aid of Wade-like rules of the sort derived from boron polyhedra. Lithium examples again present difficulties. Electrical properties have evidently not been measured to confirm that the phases are semiconductors. The more condensed clathrate-like structures such $M_8{}^1M'_{46}$, M' = Si, Ge, and Sn, as well as the remarkable infinite-bonded Li_7Ge_{12} do not appear appropriate for the present consideration.

Finally, it is appropriate to note that few isolated intermetallic phases do appear to contain relatives of the square planar Bi_4^{2-} Zintl anion. The compound $Ca_{11}Bi_{10}$ contains Bi_4 groups (D_{4h}, point symmetry), Bi_2 dimers, and isolated bismuth atoms in a ratio of 1:4:8, and the Ca-Sb, Yb-Sb and Ho-Ge systems contain isostructural phases. The symmetry relationships among these bismuth units and the calcium atoms have led to a speculative assignment of valence electrons to the clusters together with considerable delocalization between the groups.

For example, the dimers that lie above and below and end-on to each Bi_4 square unit and at somewhat longer distances appear to be bound in a n fashion. The structure is quite complex for a rigorous bonding study, but what seems important in this and other cases is that the highest energy electrons that may be more-or-less delocalized form such a cluster derived in the limit of complete charge transfer are often more nonbonding in character and what are retained to generate a recognizable cluster are the lower lying electrons in σ bonds,

A simpler interaction in a related cluster has been perceived in Na_3g_2. Here planar mercury units with D_{2h}, and nearly D_{4h} symmetry are bridged by two groups of sodium ions that naturally exhibit the same symmetry. MO calculation show that the limiting species Hg_4^{6-} is closed-shell, although, such an extreme electron transfer is, of course, quiet unreasonable. However, the sodium set with the same symmetry as the highest occupied molecular orbital gives a perfect means for delocalization of some charge back into the cations and thus, into larger arrays while still retaining some o-bonding density in the cluster. The same MO scheme naturally also applies to Bi_4^{2-} and Te_4^{2+} on the addition of either more electrons necessary to achieve close to electronetutrality. Mercury, a kind of the magic metal when it comes to bonding, forms other compounds with that alkali metals with a variety of unprecedented polymercury species, for example, infinite ribbons in NaHg but rhomboids joined at opposite corners into chains in KHg.

POLYATOMIC ZINTL ANION

Some p-block metals and metalloids react with alkali metals to form compounds such as K_2Pb_5. These compounds were investigated by Eduard Zintl in Germany in the 1930s, who found that some could be dissolved in liquid ammonia, a hospitable solvent to strong reducing agents such as these. More recently, John Corbett in the United States discovered that by complexing the alkali metal cation with 2,2,2-crypt, it was possible to prepare species in ethylenediamine solution that can be crystallized and investigated by X-ray diffraction. The crypt ligand encapsulates the alkali metal ion and thus, creates a large cation complex

$$K_2Pb_5 + 2\text{crypt} \xrightarrow{50°C} [\text{Kcrypt}]_2\,[Pb_5]$$

The stability of the product stems from the stabilizing effect that large cations have on large anions. Similar techniques have led to a variety of reduced species that have been characterized in the solid state by X-ray diffraction and in some cases in solution by NMR.

The reduction by alkali metals of Ge, Sn and Pb salts in liquid ammonia or in the presence of hexamethylphosphoramide or cryptands has allowed the isolation of many stable anions such as Ge_9^{4-}, Sn_5^{2-}, Sn_9^{3-} Pb_5^{2-}, and mixed ions of various sorts such as $Sn_2Bi_2^{2-}$, $SnTe_4^{4-}$. It may be noted that the M_4^{4-} ions are 20e species and isoelectronic with P_4, As_4 or Sb_4 and are similarly tetrahedral. The anion in $(K\ \text{crypt})_3Sn_9^{3-}$ has a tricapped trigonal prismatic structure and is paramagnetic with 21 skeletal electrons.

The Group VA elements all react with alkali and alkaline earth elements to form a considerable number of these which contain E_n^{m-} anions. Characterization of the E_n^{m-} ions has for the most part been accomplished by dissolving the solids in liquid NH_3 or $H_2NCH_2CH_2NH_2$ and adding 2,2,2-trypt to complex the cations. The resulting crystalline materials, such as $[K(2,2,2\text{-crypt})]_2Bi_4$, can then be examined by X-ray crystallography. In this way, it has been shown that the following principal types of ZintI anions exist.

1. Square E_4^{3-} for As, Sb, and Bi
2. The E_7^{3-} ions. Note that P_4S_3 is isostructural and isoelectronic with P_7^{3-}. The Bi_7^{3-} ion, though reported, may not exist.
3. As_5^{3-}, As_6^{4-}, P_{11}^{3-}, and As_{11}^{3-} are all known.

It has recently been shown that structural information can be obtained directly on solutions of the Zintl phases by EXAFS (extended X-ray absorption fine structure).

The electron-counting correlations (Wade's rules) introduced for boranes and carboranes are successful with many of these naked p-block cluster anions, and we can predict the shapes of deltahedra. As an example, the number of valence electrons available in the Pb_5^{2-} ion is $5 \times 4 = 20$ from the Pb atoms and 2 more from the charge, giving 22 electrons, or 11 pairs. Of these 11 pairs, one pair on each Pb atom, and so 5 pairs in all, are unavailable for bonding since, like the electron pair of a B-H unit in boranes, they are assumed to be directed away from the skeleton (I). Thus, the total number of skeletal electron pairs is (11–5) = 6. This number agrees with the count expected for a 5-atom (n = 5) *closo* cluster, for 5 + 1 = 6.

Example:

Correlating the electron count and structure for Zintl clusters

Determine the electron count and the structural classification for (a) the Sn_9^{4-} *ion and (b) the diamond shaped* Bi_4^{2-} *ion.*

Solution:

(a) The Sn_9^{4-} ion contains 4 × 9 + 4 = 40 valence electrons, and hence, 20 pairs. Subtracting one nonbonding pair on each atom (a total of 9) leaves 11 pairs. This fits the number expected for a nido cluster. In agreement with this conclusion, the structure is a truncated M_{10} deltahedron (II) characteristic of a *nido* cluster.

(b) The Bi_4^{2-} anion contains (4 × 5) + 2–(4 × 2) = 14 skeletal electrons, or 7 pairs. The count suggests an *arachno* cluster, but it is square planar (III) rather than the expected butterfly :

The In + 2 rule appears to be generally applicable to clusters containing main group elements. In these clusters, atoms which have no substituents are assumed to have one lone pair of electrons each, directed outward from the surfaces of the clusters. For example, Ge_9^{2-} has a tricapped trigonal prism structure, a close structure corresponding to 20 framework electrons. On the other hand Ge_9^{4-} has a nido configuration based on a bicapped archimedean antiprism.

The cation Te_6^{4+}, with 20 framework electrons, has a trigonal prism structure, which may be considered a hypho structure related to the tricapped trigonal prism structure that would presumably be adopted by hypothetical Te_9^{16+}. The rather complicated structure of the As_{11}^{3-} anion with 36 framework electrons) has been rationalized by looking upon it as a 17-vertex polyhedron from which 6 vertices have been deleted. The electronic regularities and classifications in some of these are more difficult to perceive, and the novelty and prospects for interesting science are correspondingly larger.

Background

The first evidence for what are called Zintl ions was reported by Joannis in 1891 from studies of the reaction of sodium in liquid ammonia with a variety of metals. He observed that the reaction of the characteristic blue solution of sodium with excess lead gave new solids which coloured the ammonia an intense green.

More attention was given to the precipitates which were found to have the compositions $NaPb_4$ or $NaPb_2$ depending on conditions. A brown solution from the reaction of sodium with excess antimony was also noted.

Kraus was the first to study the green sodium-lead solution, showing that it was an electrolyte with the green colour associated with the anion. He also observed that the lead solute would plate out on an anode and form at a lead cathode, that the lead solute could be precipitated on addition of normal lead (II) salts, and that tin showed similar reactions with sodium solutions. There appeared to be about two lead atoms per charge according to Faraday's law studies, leading to the formulation $NaPb_2$ for the solute.

Later, more thorough measurements by Smyth showed that an average of 2.26 Pb per Faraday plated out on the anode from the saturated solution. Similarly, a maximum of 2.24 lead atoms per Na was found to dissolve in NH_3 solutions, which was interpreted in terms of an equilibrium between $NaPb_2$ and $NaPb_3$ solutes.

A related study by Beck yielded maximum solubilities of 4.0 Te/Na and 2.33 Sb/Na, the latter value decreasing appreciably on dilution (probably because of the presence of two species). These observations led Kraus to speculate that the polyanionic salts in equilibrium with the metals (assuming single solute species) were $Na_4Pb.Pb_8$, $Na_{3-}Sb.Sb_6$, and Na_2 Te–Te_3 and that as electrolytes they contained the anions Pb_9^{4-}, Sb_7^{3-} and Te_4^{2-}, respectively.

The next information on these anions came from extensive systematic studies by Zintl and co-workers that were published in the 1930s. Zintl recognized that these solutions came about for what could be viewed as "polar" intermetallics, the more electronegative members of these lying close to the metalloid elements. X-ray diffraction was becoming available at this time as an invaluable aid for sorting phases as well as for structural investigations of the intermetallic phases.

Zintl was also a pioneer in determining the structures and properties of such polar intermetallics, which have been subsequently labelled "Zintl phases" and so the solution of many of the same materials in liquid ammonia seemed particularly relevant to him. However, subsequent research by Zintl and, more so, by others has shown that virtually none of the clusters or groupings later postulated or isolated as Zintl ions are discernible in the structures of appropriate intermetallic phases although the M^1Sn and M_1Pb phases were later found to contain tetrahedra of the heavy metal.

The first investigations reported by Zintl and co-workers were potentiometric titrations; conductance measurements were also employed in a few cases but these were relatively uninformative. Since reactions of the elements with sodium solutions were sometimes relatively slow, they instead titrated a sodium solution in NH_3 with one of a normal-valent salt of the metal. Thus, the first PbI_2 added would be reduced to the metal which then would redissolve (be reduced) relatively rapidly. The potentiometric measurements thus, followed the measured oxidation of the sodium and subsequently, of any polylead anions until a permanent precipitate of lead occurred. The reaction in the earlier stages could be described as

$$(4 + 2x)\ Na + xPbI_2 \rightarrow Na_4Pb_x + 2xNaI$$

A reference electrode was usually established early in the oxidation sequence by filling a relatively isolated compartment with a small amount of the solution.

The course of two potentiometric and one conductometric titrations in the lead system is reproduced from Zintl's original article. The compositions of the indicated ions were generally quite well marked by the data, *i.e.,* in four experiments they obtained x = 6.89 and 7.02 for Pb_7^{4-} and x = 8.86 – 9.11 (average 8.96) for Pb_9^{4-}. The former composition was seen only with the more dilute sodium solutions, as might be expected for a concentration-dependent equilibrium. However, ion pairing may have a significant effect as well since the more reduced ion was not seen in a comparable study with potassium. Stable emfs could not be obtained with tin solutions. Similar experiments with Cu, Ag, Au, Zn. Cd, Hg, or Tl iodides in ammonia solution indicated the absence of a significant anion solution chemistry for these elements. Dullenkopfalso investigated the direct solution of binary alloys and their mixtures into ammonia. Extraction of compositions (Sn, Pb)/Na = 2.25 to ~ 2.75 gave evidently pure solutions and excellent analyses for the indicated homoatomic M_9^{4-} compositions. Colloidal particles were also seen in the solutions, presumably from alloy disintegration or peptization.

Table 4 : Polyanions Deduced by Zintl in Liquid Ammonia.

	As_3^{3-} yellow	Se_2^{2-} bright red
	As_5^{3-} dark red-brown	Se_3^{2-} wine red
	As_7^{3-}	Se_4^{2-} green $Se_{5,6}^{2-}$ red-green (dichroic)
Sn_9^{4-} blood red	Sb_3^{3-} deep red	Te_3^{2-} violet
	Sb_7^{3-} red-brown	Te_3^{2-} Te_4^{2-} dark red
Pb_7^{4-} green	Bi_3^{3-} deep violet	
Pb_9^{4-} green (to violet)	Bi_5^{3-} brown	
	Bi_7^{3-} red brown	

The composition NaSb readily yielded a saturated solution analysing $NaSb_{2.33}$ ($Na_3Sb_{6.99}$) while the composition $NaBi_{1.95}$ (= $Na_3Bi_{5.85}$) was obtained only after extraction of a $NaBi_3$ alloy for several months. It was presumed that the latter represented an unsaturated solution, and so the result was rounded up to Na_3Bi_7 (rather than the closer Na_3Bi_6) for the following reasons: The composition Na_3Bi_5 had already been deduced by titration experiments described above, all the other main group 5 anions appeared to have a charge of 8-(or a multiple thereof), basically the 8-N rule (below), and all of the polyanions they had found appeared to contain an odd number of heavy atoms. As will be seen later, present evidence suggests this train of reasoning was probably in error and that the correct species may have been $Na_2Bi_{3.9} \simeq NazB_2$.

The ions so deduced in liquid ammonia by these investigators are summarized in Table 4 along with the indicated colours except that the less relevant polysulfide results have been omitted. In addition, the unlisted valence compounds Na_3MV and Na_2M^{VI} usually precipitated in the early stages of oxidation of the sodium solutions. The violet colour noted with what was presumed to be Pb_9^{4-} pertained to the rubidium and cesium systems, perhaps another hint of ion-pairing effects.

The Pb_9^{4-} species produced electrolytically was observed to be unstable when a Ei_4NI electrolyte was used. None of the characteristic green colour of this ion was observed with a Sri; electrolyte or on extraction of a $CaPb_3$ composition. Although the Sn_9^{4-} solutions could not be readily achieved by electrolysis, this and the lead analogue were also obtained in a disproportionation reaction following Bergstrom, viz.

$$11Sn + 6KNH_2 \xrightarrow{NH_3} K_4Sa_9 + 2KSn(NH_2)_3$$

Germanium gave only negative results in all of these experiments although extraction of an NaGe composition for several days gave a faint brown colour in the ammonia and a qualitative test for that element in the brown residue. Corresponding reactions with Na-Ga, Na-In, Cs-Ga, Cs-In, and Cs-Tl alloys yielded negative results. Evaporations of the metal-saturated poly lead, tin, antimony, and bismuth solution at low temperatures all gave dark, non-metallic-appearing and evidently amorphous solids that were apparently the $Na(NH_3)_n^+$ salts of the indicated anions. Removal of NH_3 by pumping went through several intermediate ammoniates to give gray, often pyrophoric, products that eventually took on a metallic reflectance. Two ammoniates were specifically shown to be amorphous to X-rays while the ammonia-free products in all cases were clearly the appropriate binary metal compounds or then-mixtures. Typical decompositions could thus, be described as

$$(Na(NH_3)_n^+)\ Pb_9^{4-} \rightarrow 4NaPb_{2.25} + 4nNH_3$$

$$\text{or}\quad [Na(NH_3)_n^+]_3\ Sb_7^{3-} \rightarrow 3NaSb + 4Sb + 3nNH_3$$

where the indicated Na-Pb composition lies within a homogeneity range of the $NaPb_{3-x}$ (Cu_3Au–type) phase. The production of only known and unrelated intermetallic phases by this route is general. The solution compositions Na_3Sb_3 and Na_3Bi_3 are of course of the same as the one-to-one phases known in the binary intermetallic systems but these must be different; NaSb contains infinite helical chains of antimony that are isoelectronic with elemental tellurium.

The bonding interpretations purposed for these anion structures at that time are naturally no longer very appropriate. Analogies with the polysulfides were strongly featured as well as the fact that the apparent charges on all of these (Table 5) appeared to follow them 8-N rule for the corresponding hydrides. The solutes were accordingly formulated as adducts of the sodium salts of the hydrides, *e.g.*, $Na_4Pb.Pb_8$, $Na_3Bi.Bi_4$, and $Na_22Te.–Te_3$, analogous to Kraus interpretations. The relationship of these phases to a salt-metal transition seen in the solid state in some binary systems as a function of composition also received considerable attention.

5 : Homopolyatomic Zintl Anions of Known Structure.

Anion	*Symmetry*	*Characteristics*	*p-electrons*	*Isoelectronic analogues*
Ge_9^{2-}	C_{2v} ($\sim D_{3n}$)	dark red rods	20	$B_9H_9^{2-}$
Ge_9^{4-}	C_{4v}		22	Bi_9^{5} + (D_{3h})
Sn_9^{4-}	C_{4v}	dark red rods	22	Bi_9^{5+}" (D_{3h})
Sn_9^{3-}	D_{3h}	dark red plates	21	
Sn_5^{2-}	D_{3h}	orange-brown plates	12	
Pb_5^{2-}	D_{3h}	ruby red prisms	12	Bi_5^{3+}
Sb_4^{2-}	D_{4h}	dark red wedges	14	
Bi_4^{2-}	D_{4h}	dark green prisms or plates	14	Te_4^{2+}, Se_4^{2+}
As_{11}^{3-}	$\sim D_3$	deep red rods	36	P_{11}^{3-}
Sb_7^{3-}	C_{3v}	brown prisms, rods	24	As_7^{3-}

The literature suggests there was negligible subsequent activity in this area for nearly four decades. Of course, negative results are seldom published. An apparently related phenomenology concerned the solution of some of the same intermetallic phases in molten salts. This behaviour was first studied in Zintl's time in NaBr by Heyman and Weber and more recently in the lower melting NaCl-NaI eutectic by Okada *et al*. The latter found that sodium alloys of Au, Pb, and Sn are measurably but only slightly soluble, usually ≤ 1 mol %, while the corresponding antimony and bismuth phases are notably more soluble.

Solutions of the less relevant valence compounds Na_3Sb and Na_3Bi cryoscopically appear to represent the simple salts although the latter is metallic in the solid state. Most interesting is the phase NaSb which dissolves to a few percent and cryoscopically behaves as Sb_3^{3-} quite well and hence, as a Zintl ion (Table 5), in contrast to the structure of the solid alloy which, again, contains infinite antimony helices. Observation of other possible Zintl anion species is presumably limited by their thermal instabilities at the melt temperatures involved (≥ 570°C in the above cases).

3

Classifiaction of Molybdenum and Walfram Elements

INTRODUCTION

Scheelite in 1778 proved that the mineral molybdenite is different from graphite as molybdenite left a white residue MoO_3 with nitric acid while graphite remained unaffected. He showed that molybdenite is a sulphide ore of molybdenum and occurs in nature widely. *Peter Jacab* in 1782 isolated the metal. The term *Molybdos* was used by the Greeks to represent certain substances like galena and other black minerals which left a mark on paper. Upto the middle of the eighteenth century great confusion remained as it was not distinguished from black lead graphite. All the three minerals like molybdnite, galena and graphite, left a black mark on paper.

Molybdenum never occurs in native state. Its important ores are :

1. Molybdenite – MoS_2
2. Wulfenite – $PbMoO_4$
3. Molybdite – FeO_3. $MoO_3.H_2O$
4. Ilsemannite – $MoO_2.5MoO_3$

Extraction

Molybdenum is commonly extracted from molybdenite and wulfenite.

From Molybdenite:

Method I

The following steps are involved :

1. **Crushing :** The ore is first of all crushed into fine powder.
2. **Concentration :** The crushed ore is concentrated by froth floation process.

   The well powdered sulphide or is placed along with water and pine oil in a tank and is agitated by a blast or air. The particles of the ore are preferentially wetted by oil and therefore, rise to the surface in the form of a scum of froth while impurities are wetted to get ore particles.
3. **Roasting :** The crushed and concentrated ore of molybdenite is heated strongly in a current of air (Roasting) when molybdenum trioxide is formed.

$$2MoS_2 + 7O_2 \rightarrow 2MoO_3 + 4SO_2$$

Method II

The concentrated ore is mixed with sodium carbonate and the mixture is fused in a reverberatory furnace in presence of excess of air,

By this operation molybdenum sulphide is converted into sodium molybdate.

$$MoS_2 + 3Na_2CO_3 + 9/2O_2 \rightarrow Na_2MoO_4 + 2Na_2SO_4 + 3CO_2$$

or

$$2MoS_2 + 6Na_2CO_3 + 9O_2 \rightarrow 2Na_2MoO_4 + 4Na_2SO_4 + 6CO_2$$

The fused mass is extracted with water. The solution on acidification by the addition of hydrochloric acid precipitates out MoO_3

$$Na_2MoO_4 + 2HCl \rightarrow MoO_3 + 2NaCl + H_2O$$

4. Purification of MoO_2 :

   (a) **Treatment with ammonia :** Impure molybdenum trioxide is dissolved in ammonia when ammonium molybdate is formed.

   $MoO_3 + 2NH_3 + H_2O \rightarrow (NH_4)_2MoO_4$

   (b) **Purification of ammonium molybdate :** It is purified by crystallisation or some other suitable method:

   (c) **Decomposition :** Ammonium molybdate is heated to decompose it to form its trioxide.

$$(NH_4)_2MoO_4 \rightarrow MoO_3 + 2NH_3 + H_2O$$

From Wulfenite

1. **Ullik's Method :** The wulfenite ore is washed with dilute HCl to wash out the oxides, carbonates and phosphates of Ca, Mn, Fe, etc and the washed material is then digested with hot conc. HCl and filtered. The filtrate is now concentrated and cooled; crystals of $PbCl_2$ which are obtained during cooling are separated and filtrate is treated with H_2SO_4 in order to isolate remaining lead salt as $PbSO_4$. The filtrate is then with NH_4OH, concentrated and cooled. During cooling the crystals of ammonium molybdate separate out. On heating, it yields MoO_3.

$$PbMoO_4 + 2HCl \rightarrow PbCl_2 + H_2MoO_4$$

$$H_2MoO_4 + 2NH_4OH \rightarrow (NH_4)_2MoO_4 + 2H_2O$$

$$(NH_4)_2MoO_4 \rightarrow MoO_3 + 2NH_3 + H_2O$$

2. **Modified Method of Wohler :** It involves the following steps.

   (i) The concentrated ore is fused with sodium sulphide and sulphur.

$$PbMoO_4 + 3Na_2S + 3S \rightarrow Na_2[MoS_4] + PbS + 2Na_2O + SO_2$$

   (ii) After cooling the solid mass is dissolved in water whereby sulphides of sodium and calcium along with sodium thiomolybdate go into solution.

   (iii) The filtrate is acidified with H_2SO_4 so that molybdenum trisulphide is precipitated.

$$Na_2[MoS_4] + H_2SO_4 \rightarrow MoS_3 + Na_2SO_4 + H_2S$$

   (iv) The precipitate is filtered, washed with water, dried and heated in absence of air. It is roasted in air to molybdic oxide.

$$MoS_3 \rightarrow MoS_2 + S$$

$$MoS_2\ 7O_2 \rightarrow 2MoO_3 + 4SO_2$$

3. **Modern Method :** The mixture of concentrated wulfenite ore and Na_2CO_3 is fused. The formation of sodium molybdate takes place.

$$Na_2MoO_4 + 2HNO_3 \rightarrow H_2 + 2NaNO_3$$

The free molybdic acid is converted to ammonium molybdate which on heating yields MoO_3

$$H_2\,MoO_4 + 2NH_4OH \rightarrow (NH_4)MoO_4 + 2H_2O$$
$$\downarrow \text{on heating}$$
$$2NH_3 + MoO_3 + H_2O.$$

Metallurgy

Molybdenum metal may be prepared in several ways :

(i) By reducing the oxide, MoO_3 in an atmosphere of hydrogen.

$$MoO_3 + 3H_2 \xrightarrow{500\text{-}1200°C} MO + 3H_2O$$

Molydbenum is obtained as a fine dark powder.

(ii) **Alumino-thermic process :** Alumino thermic reduction of MoO_3 has been used for preparing metallic molybdenum.

$$MoO_3 + 2Al \rightarrow Mo + Al_2O_3$$

By this method the metal obtained is 98–99 percent pure containing iron and silica as impurities, but is free from carbon.

(iii) 99 percent pure Mo can be prepared by the reduction of MoO_3 with calcium or zinc dust.

(iv) By heating a mixture of molybdenum sulphide and dioxide in an electric furnace, molybdenum of 99.5 percent purity having 0.5 per cent of sulphur, can be obtained.

In all the methods described above molydenum is obtained in the form of a powder. The metal is obtained in the compact form by pressing it at temp. near its melting point. This heating is done in an atmosphere of H_2.

Properties

Physical

1. Pure molybdenum is a silvery white metal but obtained as a metallic power. It resembles tugsten in some of physical properties.
2. It is a quite malleable, ductile, and softer than steel, but impurities make it brittle. Pure metal can be polished easily. It can be drawn into wires, ribbons and thin sheets.
3. Its specific gravity is 10.2.
4. Its melting point lies between 2620°C and 2680°C and boiling point 3000°C.

Chemical : Molybdenum behaves as an active or noble metal.

1. **Action of air :** Molybdenum oxides slowly in air at ordinary temperature but burns in oxygen at 400–600°C forming oxide MoO_3.

$$2Mo + 3O_2 \rightarrow 2MoO_3$$

2. **Action of nitrogen :** Nitrogen is slightly absorbed by molybdenum wire or plate. Unannealed quantity of molybdenum wire absorbs considerably nitrogen at elevated temperature and forms an unstable nitride.

3. **Action of water :** Pure molybdenum is not attacked by water, but due to slow action, a blue film is formed on the metal.

4. **Action of H_2O_2 :** The metal is slightly oxidised by hydrogen peroxide.

5. **Action of halogens :**

   (a) Molybdenum reacts with fluorine at ordinary temperature forming a volatile fluoride.

$$Mo + 3F_2 \rightarrow MoF_6$$

   (b) Chlorine reacts at dull red heat.

$$Mo + 2Cl_2 \rightarrow MoCl_4$$

   (c) Bromine reacts at bright red heat.

$$Mo + 2Be_2 \rightarrow MoBr_4$$

   (d) Iodine does not react even up to a temperature of 800°C.

6. **Action of halogen acids :** Molybdenum does not react with hydrofluoric acid or hydrochloric acid.

7. **Action of sulphuric acid :**

   (a) *Dilute sulphuric acid has no action on the metal.*

   (b) *Hot concentrated sulphuric acid attacks that metal.*

$$3H_2SO_4 + \mathit{Mo} \rightarrow MoO_3 + 3H_2O + 3SO_2$$

8. **Action of nitric acid :** Hot dilute nitric acid dissolves the metal.

$$6HNO_3Mo \rightarrow MoO_3 + 3H_2O + 6NO_2$$

9. **Action of aqua-regia :** This dissolves the metal forming a blue solution.

10. **Action of fused alkali :** Fused alkali attacks it forming alkali molybdate.

$$8NaOH + 2Mo \rightarrow 4Na_2MoO_4 + 4H_2$$

11. **Action of non-metals :** Non-metals like sulphur, phosphorus, boron, carbon and silicon combine with molybdenum.

12. **Action of metals :** Molybdenum alloys with several metals chiefly iron and forms molybdenum steel having several useful properties. Molybdenum alloys with nickel and forms elongation resistant alloy.

Uses

It finds wide range of applications.

1. It is used as an anticathode in X-ray tubes.
2. With tungsten, molybdenum is used as thermocouple for high temperature measurements.
3. Molybdenum is used in making special steels which are used for making rifle barrel, high speed lathe tools.
4. Molybdenum is small amounts (2–4 %) to stainless steel to enhance resistance to corrosion.
5. In the manufacture of standards weights.
6. Being cheaper than platinum it is used as a winding for electric resistance in resistance furnaces.
7. As promoter in Haber's process for the manufacture of ammonia.

$$N_2 + 3H_2 \xrightleftharpoons{Fe/MO} 2NH_3$$

8. Ammonium molybdate is used as laboratory reagent for the detection and determination of phosphate and arsenate.
9. Ammonium molybdate has been used in fire proofing and as a bactericidal agent.
10. Phosphomolybdic acid is used as a reagent for the precipitation of alkaloids and albumin in water solutions.

History and Nomenclature

A new era in the coordination chemistry of alkali elements was inaugurated by the discovery of new types of ligands, macrocyclic-polyethers ("crown") by Pedersen in 1967. Pedersen reported syntheses of over 50 cyclic polyethers in which the size of the macrocyclic rings, the number of ether oxygens, and the number and type of substituent groups on the ring varied widely. For example, the polyether rings contained from 9 to 60 atoms, of which 3 to 20 were oxygen atoms. Structures of some typical crown ethers, their trivial names and code designation.

1.	Dibenzo-14-crown-4	DB14C4
2.	Dicyclohexyl-14-crown-4	DC14C4
3.	Cyclohexyl-15-crown-5	C15C5
4.	18-crown-6	18C6
5.	Dibenzo-18-crown-6	DB 18C6
6.	Dicyclohexyl-18-crown-6	DC18C6
7.	4,4'-dimethyldibenzo-18-crown-6	4,4'DiMcDB19C6
8.	Tribenzo-21-crown-7	TB21 C7
9.	Dibenzo-30-crown-10	DC30C10
10.	Dicyclohexyl-60-crown-20	DC60C20.

Pedersen suggested useful and descriptive trivial names for these fascinating compounds. He proposed that the cyclic polyethers be called "crowns" and be identified by (a) the number and kind of substituent groups on the ring, (b) the total number of atoms in the ring, (c) the class name-crowns, and (d) the number of oxygens in the ring. Thus, the trivial name of compound 5, is dibenzo-18-crown-6, while according to the IUPAC nomenclature it is called 2,3,11,12, –dibenzo– 1,4,7, 10,13,16-hexaoxa cyclooctadeca-2,11–diene.

It was immediately recognized that these compounds have an unusual ability to form relatively strong complexes with alkali and alkaline earth ions. In fact, in the presence of some of these ligands alkali salts can be solubilized in nonpolar organic solvents. Perhaps the most spectacular effect was the dissolution of potassium permanganate in benzene by an 18-member polyether ring DC18C6.

11.	$a = b = c = 0$	Cryptand 111	C111
12.	$a = b = 0, c = 1$	Cryptand 211	C211
13.	$a = 0, b = c = 1$	Cryptand 221	C221

14. a = b = c = 1	Cryptand 222	C222
15. a = b = l, c = 2	Cryptand 322	C322
16. a = l, b = c = 2	Cryptand 332	C332
17. a = b = c = 2	Cryptand 333	C333
18. –	Monobenzo cryptand 222	C222B
19. –	Dibenzo cryptand 222	C222Bz

The discovery of the crown ethers was soon followed by synthesis of macrobicyclic polyethers containing three polyether strands joined by two bridgehead nitrogens. These compunds have three-dimensional cavities which can accommodate a metal ion of suitable size and form an inclusion complex.. Once again the IUPAC names for these compounds are quite cumbersome. Since it has been shown that complexation by a bicyclic ligand involves envelopment of the metal ion these ligands are called [2] *cryptands* [2] indicating a bicyclic ligand), and the corresponding complexes, *cryptates*.

Since bicyclic ligands may have a variable number of ether oxygens in each strand it is convenient to use the number and the distribution of oxygen atoms for coding the cryptands. As shown in Fig. 5.5, a compound with two oxygen atoms on each hydrocarbon strand is cryptand 222, hereafter abbreviated C 222. The IUPAC name for this compound is 4, 7, 13, 16, 21, 24, hexaoxa-1, 10-diazabicyclic (8, 5, 8) hexacosane. Trivial names and code designations of [2] cryptands. Similar ligands containing three and four macrocycles ([3]-cryptands and [4]-cryptands) have been synthesized as well. Typical structures.

22. Y = CH_2	23. Y = O	24. Y = NH
25. Y = CH_2	26. Y = O	27. X = H_2
28. X + O	30. Y = O	31. Y = H_2

Crown and cryptand ligands have been prepared in which some, or all, of the ether oxygens are replaced by nitrogen or sulphur atoms. As will be seen later, such substitutions can drastically modify the complexing abilities of the ligands. Macrobicyclic compounds with three hydrocarbon chains attached to two nitrogen bridgeheads were synthesized by Simmons and Park. The length of the chains varied from 6 to 10 carbon atoms. Finally, certain naturally occurring antibiotic ionophores such as valinomycin, monensin, nigericin, etc. have been shown to complex alkali metal ions, but this is an extensive subject in its own right and will not be discussed in this chapter, which is focused primarily on the physicochemical studies of crown and cryptate complexes with metal ions. Different topologies of the crown and cryptand ligands, where Z represent bridgehead moieties.

Crown ethers, schematically represented by A, have two-dimensional cavities that can accommodate an ion with spherical symmetry of proper dimension. The [2]- cryptand, B, as well as [3 [-cryptands, C (cylindrical) and D (spherical), and the [4]-cryptand, E, all have three-dimensional cavities that are capable of forming an inclusion complex with a cation (in case D, also with an anion) or cations of proper size. The task of presenting an adequate or even semi adequate description of the current state of this field of research is made very difficult by the rapid growth in the popularity of macrocyclic ligands. Numerous reviews have been published within the last 5 years. However, advances in this field are so prolific and rapid that the reviews are outdated by the time of publication. This chapter is not an exception.

SYNTHETIC METHODS

I. Crown Polyethers

Aromatic crown polyethers were prepared by Pedersen by condensation reactions of a bis-pheaol with a dichloride. A typical reaction is given below: The yield strongly depends on the nature of the cation M^+ which seems to indicate that the condensation is facilitated by the template effect of M^+. Thus for example, if potassium hydroxide is used instead of sodium hydroxide the yield of DB18C6 is substantly increased. On the other hand, a base with a large organic cation, n-Bu_4NOH, drastically decreased the yield. Potassium ion fits quite comfortably into the DB18C6 ring, while Na^+ ion is too small and n-Bu_4 N^+ ion is too large.

Catalytic hydrogenation of aromatic hydrocarbons yields the corresponding saturated polyethers. The unsubtituted polyether 18C6 was prepared by condensation of hexaethylene glycol with 1, 2-dimethoxyethane in the presence of potassium t-butoxide. Numerous authors followed Pedersen's work in developing new synthetic paths for crown ethers. For example, an improved synthesis of 18C6 is suggested by Gokel *et al.*, who reacted triethylene glycol with triethylene glycol dichloride in 10% aqueous tetrahydrofuran. It is interesting to note that this crown easily forms a crystalline adduct with acetonitrile. Under gentle heat and high vacuum acetonitrile is volatilized and pure crown 18C6 is obtained. Synthetic methods for the preparation of 12C4 and 15C5 have been developed by several workers. For example, base condensation of ethylene glycol with 1, 8- dichloro-3, 6 dioxaoctane in dimethyl-sulfoxide in the presence of lithium perchlorate gives a 13% yield of 12C4.

Numerous crown compounds have been prepared in which some of the ether oxygens have been replaced by sulphur atoms. Pedersen prepared sulphur analogs of DB15C5, B15C5, and DB18C6 containing two, and in the latter case, four sulphur atoms by condensing mercaptophenols or ditbiols with terminally substituted ether dichlorides. In more recent years numerous mixed oxa-thia crowns were synthesized by Bradshaw and co-workers by condensation of oligoethylene glycol dichloride with a dimercaptan. Synthesis of that crowns was reviewed by Bradshaw and Hui in 1974.

Crowns with mixed ether-ester and ether-ester amide functions also have been synthesized recently by Bradshaw and co-worker. Some typical examples are given below. The compounds were obtained by reacting dioxodioic acid chlorides with various oligoethylene glycols or with ethylene diamine :

Synthesis of 4'-vinyl derivatives of B15C5, B18C6, and DB19C6 was first described by Kopolow et al. A simple version of the synthesis was also published. The two first monomers can be polymerized to give linear polymers :

The average molecular weight of the polymers was found to be ~ 40,000.

II [2]-Cryptands

1. **Diazamacrobicycles :** Diazabicycloalkanes were prepared by Simmons and Park by reacting a diazacycloalkane with a diacid chloride :

   Bicyclic compounds were prepared with chains of equal length, $= l = m = 7 - 10$, as well as with chains of unequal length, [6,6,4], [6,8,10] and [8, 8,10].

2. **Diazapolyoxamacrobicycles :** The synthesis of diazapolyoxa-macrobicyclic polyethers is more involved and tedious than the synthesis of crown ethers see the reaction scheme. In the case

of cryptand 222, a monocyclic diazapolyether is first formed by a high-dilution reaction between 1,8-diamino-3,6-dioxa-octane and triglycolyl dichloride in benzene solution. The dilactam is reduced with lithium aluminium hydride in tetrahydrofuran solution. The second cyclization is carried out with another molecule of triglycolyl dichloride; the resulting dilactam is reduced with diborane to a bis-(borane-amine) adduct of C222, which is treated by 6 MHCl to yield the dihydrochloride derivative. The product is dissolved in water and passed through an anion exchange column in the hydroxide form to yield the free base C222. By using this technique and varying the number of ether oxygens, compounds have been prepared from Cl 11 to C333.

The high-dilution technique requires very gradual (~ 8 h) mixing of reagents in order to avoid polymerization reactions. It has been shown by Dye *et al.*, that the reaction can be carried out in less than one minute if the reagents are efficiently mixed in a suitable flow cell.

3. **Polythia and Polyoxa [2]-Cryptands :** By substituting a dithia-diamine or a dithia dicarboxylic acid dichloride for the corresponding oxygen species, diazaoxathia macrobicyclic cryptands were prepared. In particular, analogs of C222 were synthesized that contained two sulphur atoms in one, two, or all three strands.

Similarly, by replacing oxygen atoms by N-R (R = H, CH_3, $COOCH_3$, or Ts) groups, cryptands containing nitrogen atoms in the hydrocarbon strands were produced.

[3]-and [4]-Cryptands

Syntheses of cylindrical and spherical [3]-cryptands (compounds 21-31, 5.6) have been described in several publications. The general scheme for the stepwise synthesis of cylindrical [3] and [4]-cryptands. The reaction sequence for the synthesis of compounds 25.

METAL-CATION COMPLEXES: PREPARATION AND STRUCTURE

Monocyclic Ligands (Crowns)

In his classical paper on the synthesis and properties of cyclic polyethers, Pedersen also described the preparation of over 50 crystalline

1:1 complexes of various crowns, mostly with alkali or alkaline earth salts. In a subsequent paper synthetic methods for the crown complexes are developed in more detail and it was found that in case where the size of the ring was smaller than the size of the complexed ion, crystalline complexes were often obtained which contained two molecules of ligand per metal ion. Pedersen postulated that in such cases "sandwich" complexes are formed in which the two ligand rings are held together by a small spherical ion. Complexes with inverse "sandwich" arrangement M^+-crown- M^+ were not observed in this study however their existence was demonstrated in later investigations.

Cesium complexes with B18C6, DB18C6, and DC18C6 were also obtained with $(crown)_3$ $(CsX)_2$ stoichiometry. Extending the previous hypothesis, Pedersen thought it possible that these complex have a "club sandwich" structure Crown-Cs^+-Crown Cs^+–Crown.

In principle, the preparation of solid crown complexes with alkali salt is quite simple. The desired metal salt and the ligand are dissolved in an appropriate warm nonaqueous solvent, and on cooling, the complex precipitates out. In general, ethanol seems to be the solvent of choice although many others have been used. Obviously the formation of the crystalline complex is determined by relative solubilities of the reactants and the products in the given solvent, Equimolar mixtures of KSCN, Ca $(SCN)_2$, and BI $5C_5$ in ethylacetate give the calcium complex, while the same mixture in methanol yields the potassium complex.

In a number of cases, although reactions were carried out in "anhydrous" solvents, the complexes obtained contained one or more molecules of water of crystallization. Poonia and Truter pointed out that rigorous exclusion of water is necessary in order to obtain anhydrous complexes. Occasionally, but not frequently, a solvate other than the hydrate is formed. Parsons and Wingfield obtained calcium thiocyanate-B15C5 with methanol or pyridine of solvation when the reactions were carried out in these respective solvents.

Parsons *et al.*, prepared a series of alkali tetraphenylborate complexes of B15C5,18C6, DB18C6, DB24C8, and DB30C10. In most cases 1 : 1 complexes were obtained. However, with B18C6 sodium, potassium and cesium complexes were prepared with L_2MBPh_4, stoichiometry (L = ligand). It is interesting to note that B15C 5 gave 1 : 1 complexes only with NaBPh,, and these complexes invariable contained one molecule of solvent (water or an alcohol) in the crystal lattice.

On the other hand, with a large ligand, DB30C 10, both L. NaX and $L(NaX)_2$ crystalline complexes were obtained. In the latter case the crystals aıso contained one or two molecules of water although they were prepared in presumably anhydrous solvents, dichloromethane. Fig. 5.11 methanol, and tetrahydrofuran. Alkali complexes of cryptands also often contain some water of crystallization.

Crystal structures of a number of crown complexes have been determined by many investigators. Pioneering work in this field was done by M.R. Truter and co-worker. In general, the studies indicate that for the 1:1 complexes the cations are located at or near the centre of the oxygen atom ring and all the oxygen atoms participate in the coordination. The metal ion is also coordinated to water of hydration and/or to the anion in the direction perpendicular to the plane of the ring.

If the size of the cation is somewhat larger than the ring, the former can be situated above the plane of the oxygen atoms. Thus, in benzo-15-crown-5-sodium iodide complex, B15C5. Nal-H_2O, the cation is 75 pm out of the plane of the five oxygen atoms. It is also coordinated to the water molecule, which is hydrogen bonded to the anion. The sodium-oxygen distances of 235-243 pm are very nearly equal to the sum of the sodium and oxygen crystallographic radii.

The structure of dibenzo-18-crown-6 (DB18C6) complex with rubidium thiocyanate was determined from an analysis of a mixed- cation complex, DB18C6. $Rb_{0.55}$ $Na_{0.45}$ SCN, and it is illustrated. The Rb^+ ion- is almost coplanar with the six oxygen atom and is coordinated to the thiocyanate ion through the nitrogen. A detailed study of 18C6 and its complexes has been carried out by Dobler, Dunitz, and co-workers. In the free ligand the three OCH_2CH_2O subunits are clearly different. The mean carbon-carbon and oxygen-oxygen distances are both short compared to gaseous 1, 4-dioxane.

In the sodium complex, 18C6, NaSCN H_2O, the sodium ion is coordinated to six oxygen atoms of the ring, and to a water molecule. One oxygen atom of the ligand is out of the plane of the other five. On the other hand, the potassium thiocyanate complex 18C6.KSCN was obtained in anhydrous form; the K^+ ion occupies a crystallographic centre of symmetry and is also coordinated to six oxygen atoms of the ligand. In the case of larger alkali cations, rubidium and cesium, the cation is displaced from the mean plane of the polyetherring, by~ 120 pmfor Rb^+ and ~144 pm for Cs^+. Finally, for the calcium thiocyanate complex 18C6.

Ca $(SCN)_2$, each cation is surrounded by a nearly planar hexagon of oxygen atoms.

A crystalline complex of magnesium (II) chloride with a small crown, 12 C4, with the composition $MgCl_2$. 12C4.6H_2O, has an unusual structure. The metal ion is not found in, or near, the macrocyclic ring. Instead the ligand is hydrogen bonded to the water molecules coordinated to the Mg^{2+} ion. The results indicate that there is little chance for the existence of a magnesium (II) –12C4 complex in aqueous solutions. In fact Mg^{2+} is known to have very high affinity for water and it does not readily undergo complexation reactions in aqueous solutions.

In contrast to magnesium, sodium chloride forms a sandwich complex with 12C4 with the stoichiometry NaCl.$(12C4)_2$. 5H_2O. The cation is insulated from the anion and the water molecules. A similar structure has been found for the sodium hydroxide complex NaOH $(12C4)_2$. 8H_2O. A structure intermediate between the Na^+ and Mg^{2+} cases is that of the calcium chloride complex $CaCl_2$ 12C4. 8H_2O, where the cation is octa coordinated to the four oxygen atoms of the crown on one side and four water molecules on the other.

A copper (II), chloride complex of 12C4 was obtained in anhydrous form. In this case the metal ion is hexa coordinated in the form of a distorted octahedron with the two chloride ions in cis position. Another case of crystalline crown complex in which the cation is well outside the ligand ring is that of the uranyl complex of 18C6. The compound was first prepared by Costes *et al.* On the basis of infrared data on the solid adduct $UO_2(NO_3)_2$. 18C6.4H_2O, the authors assumed that this was a *bonafide* complex with the cation in the centre of the ring.

However, a recent crystal structure determination of the dihydrate $UO_2(NO_3)_2$.18C6. 2H_2O by Bombieri et a! .indicates that the uranyl ion is coordinated to the two nitrate ions and to two oxygen atoms of the water molecules. There is no direct ligand-cation bonding, but the former is weakly hydrogen-bonded to the water molecules, which thus, form a link between it and the cation.

In the above complexes and in similar 1 : 1 crown-alkali or crown-alkaline-earth complexes, the ligand molecules remain quite inflexible as long as the size of the ring and that of the ion are commensurate. On the other hand, if the size of the ring is considerably larger than the cation, the ligand may deform so as to bring all of its donor atoms into a bonding contact with the cation.

An X-ray study of dibenzo-30-crown-10 (DB30C10) complex with potassium iodide shows that this ligand is twisted around the potassium ion ("as seems of a tennis ball") so as to bring all ten oxygen atoms into coordinating positions. The potassium-oxygen distances are in the range 285.0-293.1 pm, which again is close to the sum of the ionic radius of K^+ and the van der Waals radius of oxygen. There is no interaction between potassium ion and the iodide ion.

In the case of $(Crown)_2$ MX complexes, the X-ray investigations confirmed-Pedersen's assumption of a "sandwiched structure". For example, in the $(B15C5)_2$ KI complex the K^+ ion is sandwiched between two centrosymetrically related crown molecules, the five oxygen atoms of each ligand coordinated to the cation are approximately coplanar. The iodide ions occupy the spaces between the complexed cations at distances which preclude any cation-anion interaction.

It appears that structures 3 : 2 complexes such as $(DB\ 1806)_3$ $(CsSCN)_2$ and $(DC18C6)_3(CsI_3)_2$ have not been determined as yet and the question of the "club sandwich structure appears to still be unresolved.

Table 1 : Reported Crystalline Structures of Crown Complexes.

18C6	DC 18C6 "A"
DC18C6 "B"	12C4. $MgCl_2.6H_2O$
$(12C4)_2NaC1.5H_2O$	$(12C4)_2NaOH.8H_2O$
12C4.$CoCl_2.8H_2O$	12C4.$CuCl_2$
B15C5.NaI.H_2O	$(B15C5)_2KI$.
18C6.NaSCN.$H_{22}O$	18C6.KSCN
18C6.RbSCN	18C6.CsSCN
18C6.$Ca(SCN)_2$	18C6.Ktosylate
18C6.K ethyl acetate enolate	18C6. $UO_2(NO_3)_2.2H_2O$
DB18C6.NaBr.$2H_2O$	DB18C6.NaBr.$2H_2O$
DC18C6.NaBr.$2H_2O$	DC18C6.$Ba(SCN)_2$
DB24C8.2KSCN	DB30C10.KI
Trithio 12C4	Dithio 15C5
Dithiol8C6	Dithial8C6. $PdCl_2$

Of particular interest is the structure of the complexes of two metal salts with one crown ether. The crystal structure of *bis-*

(potassiumthiocyanate) dibenzo-24-crown-8 complex is. It is seen that the complex does not have an "inverse sandwich" structure. The two cations are accommodated within the nearly planar ring of the eight oxygen atoms. Each cation is bound to five oxygens with two oxygen atoms shared by both cations. The cations are also bridged by the nitrogen atoms of thiocyanate anion.

Table 1 lists the crown complexes whose crystal structures have been reported in the literature.

MACROPLYCYCLIC LIGANDS (CRYPTANDS)

Macrobicyclic Complexes-[2]-Cryptates

Pioneering work on the crystal structure of numerous diazapolyoxa macrobicyclic complexes ([2]- cryptates) has been performed by R. Weiss and co-workers. In several cases, as with crowns, crystalline complexes contain water of crystallization. Proton magnetic resonance of diazabicycloalkanes $N\{(CH_2)_{k,l,m}]_3N$ in solutions by Simmons and Park showed that the molecules can exist in three different configurations, the *out-out* (o, o) in which the lone pair of the nitrogen atoms are directed away from the central cavity, the out-in (o, i) and the in-in (i, i) where the lone pairs are inside the cavity. These configurations are shown schematically they have also been designated as exo-exo, *exo-endo*, and endo-endo respectively. Structures of the uncomplexed ligand C222 and of its *bis*-borohydride derivative have been determined by Metz *et al.*

The free ligand has the *endo-endo* conformation similar to that of the complexed ligand (see below). However, the conformation of the three bridges (*e.g.*, around the central C-C bond) is different from that found in the complexed ligand. It should be remembered that the observed conformation in the crystalline state may be quite different from the one in solution.

The bis-(borane-amine) derivative, on the other hand, is *exo-exo* with the borohydride groups directed outside of the ligand cavity. The structure of the smallest bicyclic cryptand C111 in the form of its mono (borane-amide), C111.BH_3, has been recently determined by Weiss and coworkers. As expected, the ligand is in the exo-endo configuration with the BH_3 group being outside the cavity. The N-N distance is 494 pm as compared with 687 pm in uncomplexed C222. The cavity of C111 is small and only Li^+ has been found to form a complex. This ligand also forms efficiently shielded proton cryptates whose deprotonation kinetics are very slow.

The metal cryptates in general have 1:1 stoichiometry and the ligand is in the endo-endo configuration. Alkali complexes of the cryptand C222 present a particularly interesting pattern in that the ligand shows considerable flexibility and can adapt its conformation to accommodate alkali ions from sodium through cesium. The crystal structures of potassium, rubidium, and cesium complexes with cryptand-C222 were determined by Weiss *et al.* Potassium iodide and rubidium and cesium thiocyanates were used as the alkali salts. In the two latter cases the crystals also contained one molecule of water of hydration.

The crystals belong to the monoclinic system and contain four molecules of the complex per unit cell. The cations are located in the centre of the ligand cavity and are surrounded symmetrically by the eight heteroatoms of the ligand. The latter is in the *endo-endo* conformation. The cation-anion distances are quite long (695 pm in the case of the KI complex). It is evident, therefore, that there is no cation-anion Fig. 5.17. contact in the solid complex.

In the case of the potassium complex the cation-oxygen and cation-nitrogen distances are equal to the sum of the van der Waals radii of the atoms and the ionic radius of K^+. In the other complexes the distances are somewhat shorter. The study shows that the cryptand cavity can adapt, within certain limits to the ionic radius of the enclosed cation. The size of the cavity of C222 is close to that of the potassium and rubidium cations. Thus, the formation of potassium and rubidium cryptates occurs without undue distortion of the ligand. On the other hand, introduction of cesium ion into the C222 cavity can only be accomplished by the adaptation of the ligand to a larger cation.

For example, for the K^+, Rb^+, and Cs^+ complexes we have the following variations in bond lengths and bond angles : N-N distances - 575,600, and 607 pm; O-O distances (for atoms on different chains)- 426, 239, and 450 pm; the N-C-C-O dihedral angles are 54°, 57° and 70°, respectively. These data indicate that the stabilities of the three complexes should be in the order $K^+ > Rb^+ > Cs^+$. This is indeed the order of the respectively stability constants found in solutions.

The heats of formation are in the order $Rb^+ \sim K^+ >$ Cs. The crystal structure of the sodium-iodide –C222 complex, in general is very similar to that of other alkali cryptates except that the Na-N and Na-O bond distances of the 275 and 257 pm are considerably larger than the sum of the Na^+ ionic radius and the van der Waals radii of N and O, respectively (252 and 242 pm).

The ligand must undergo a considerable deformation in order to accommodate small Na^+ small icns. As expected, the stability of the C222.Na^+ complex in solutions is lower than that of C222.K^+

The crystal structure of the thallium (I) formate –C222 complex was found to be very similar to that of the corresponding potassium and rubidium complexes. The Tl-O distance of 290.4 pm is very close to the sum of the ionic radius of TI^+ and of the vander waals radius for oxygen. On the other hand, the TI-N distance is somewhat shorter than the respective sums of the two radii (294, vs. 302 pm), which could be due to a slightly covalent character of the TI-N bond. It should be noted that these results are in accord with the superior stability of the thallium (I) cryptates.

In the case of the lead thiocyanate complex, the cation again occupies the centre of the ligand cavity in the endo-endo form. The cation is bonded to the ten heteroatoms and to the two anions, one through the nitrogen and the other through the sulphur atoms. The Pb-O and Pb-N (N of the macrycycle) bonds are longer than the sum of the ionic radius of Pb^+ and van der Waals radii of heteroatoms. The cation is bonded to O, N, and S through electrostatic interactions and, as in previous cases the, ligand adapts the dimensions of the cavity to the size of the cation. The distance between the two nitrogen atoms in the complex is 576 pm as compared to 687 pm in the free ligand.

The structure of the corresponding barium complex, $Ba(SCN)_2$ .C222. H_2O, differs in that the cation interacts with the eight heteroatoms, with the water molecule and with one of the anions (through N). Since K^+ and Ba^+ ions have nearly the same ionic radii the structures of their C222 complexes are nearly the same. With a larger cryptand C322, a dihydrated complex Ba $(SCN)_2$. C322.2H_2O was obtained. In this case the cation interacts with the nine heteroatoms of the ligand, with both water molecules, but not with the anions.

Mathieu and Weiss determined the crystal structure of a cobalt (II) thiocyanate complex with C221, $[C221 : Co]^{2+} [Co(SCN)_4]^{2-}$. The cobalt (II) ion is completely enclosed in the ligand cavity and it is coordinated to the seven heteroatoms of the C221. The C-C torsion angles vary from 36° to 52° indicating that the ligand is strained. It should be noted that the size of the Co^{2+} cation is much smaller than the size of the C221 cavity. The structure of the lithium-iodide-C221 complex is similarto that found for C222 alkali cryptates. The ligand conformation is again *endo-endo* and the cation fits singly into ligand cavity.

A very different structure was found for the palladium (II) chloride complex with a ditihia analog of C222 hereafter designated $C22S_2$. In contrast to the alkali complexes, the palladium ion is *outside* the cavity and is bound, in a square-planar configuration of the two sulphur atoms and to the two chlorines. Nevertheless, the ligand is in the *endo-endo* configuration (N-N distance is 613.8 pm as in other complexes). The strand with the two sulphur atoms is different from the two other strands in that the hydrogens attached to the carbon atoms between the two sulphurs are inside the cavity and form hydrogen bonds with oxygen and nitrogen atoms.

Macrotricyclic Complexes-[3]-Cryptates

'The crystal structure of the silvernitrate complex with the [3]-cryptand, 21, was determined by Wiest and Weiss. Slow evaporation of methanol-butanol solution of the ligand of silver nitrate yielded crystals with stoichiometry $21.3AgNO_3$. Two of the silver atoms are located inside the molecular cavity of 21 with the $Ag^+ - Ag^+$ distance of 388 pm. Each silver atom is bound to five heteroatoms of the ligand and to an oxygen atom of a nitrate ion. The third silver ion is located outside the ligand and is bound to all three NO_3^- anions, forming an $[Ag(NO_3)_3]^{2-}$ unit.

Mellinger *et al.,* determined the structure of a 2 : 1 complex of 22.2 Na^+. The sodium ions are located in the two lateral cavities and each atom is bound to two nitrogen atoms and five oxygen atoms. The $Na^+–Na^+$ distance is 640 pm.

The spherically symmetrical [3]-cryptand, 29, can form both cationic and anionic complexes. In the cationic complex. $29.NH_4I.H_2O$, and the anionic complex, $29.4HC1.7H_2O$, the ligand has all four nitrogens in the endo configuration. In both cases, the ligand encapsulates the species NH_4^+ and H_4C1^{3+}. The chloride ion of the second complex is in the centre of the cavity and it is hydrogen bonded to the four protonated nitrogens of the ligand. Its structure is schematically. It is seen from the above brief discussion that one cannot speak of a "coordination number" of an alkali or an alkaline earth cation in a crown complex or in a cryptate.

The number of metal-ligand bonds varies drastically with the size of the ring or cavity, the configuration of the ligand, the number and nature of the donor atoms, and the case of access of the anion and/or the solvent molecules "of solvation" to the complexed cation.

Applications

Crown ethers and related species have numerous interesting properties that have led to many applications.

1. **Cation complexation :** The striking ability of neutral macrocyclic polyethers to complex with alkali and alkaline-earth cations as well as a variety of other species has proved of considerable interest to the chemistry community. Crown ethers may complex the cation associated with an organic salt and cause separation of the ions. In the absence of cations to neutralize them, many anions show considerably enhanced reactivity. For example, decarboxylations, rearrangements and nucleophilic substitutions all proceed at enhanced rates in the presence of crowns.

   By far, the most common effect of crown or cryptand compound is to enhance the reactivity of counteranions by complexation of cations. There are two notable exceptions. First, when complexed by crowns, the normally unstable or explosive a renediazonium cations exhibit remarkably enhanced stability. More dramatic is the complete isolation of sodium cation, which allowed the preparation of Na^- for the first time. The crystals of cryptated Na^+ in the presence of Na^- are deep gold in colour. This ability of cryptands to surround and protest the cation has also allowed the preparation of the first electride, a cation salt of a lone electron.

2. **Stabilization of unusual species :** The most remarkable example of this phenomenon is the preparation of stable salts of alkali metal anions. Thus, the salt $[(2.2.2)Na^+]$ Na^- formed gold-coloured crystals stable at room temperature, and was characterized by X-ray structural analysis and Na nuclear magnetic resonance spectroscopy . Stabilization of metal cluster anions such as Sb_73^-, Pb_5^{2-}, and Sn_9^{4-} by cryptation of the cation (Na^+) has been achieved. In the absence of the cryptand these compounds revert to an alloy of sodium and the heavy metal.

3. **Phase-transfer catalysis :** One of the important modem developments in synthetic chemistry was the use of the phase-transfer technique. Nuclophiles such as cyanide are often insoluble in media which dissolve organic compounds with which they react. Thus, 1-bromooctane may be heated in the presence of sodium cyanide for days with no product formation. When a

crown ether is added two things change. First, solubility is enhanced because the crown wraps about the cation, making it more lipopbilic. This, in turn, makes the entire salt more lipophilic. Second, by solvating is the cation, the association between cation and anion weakened, thus, activating the anion for reaction. This approach has been used to assist the dissolution of $KMnO_4$ in benzene (to give so-called purple benzene) in which solvent permanganate is a powerful oxidizing agent. One striking example of solubilization is the displacement of chloride by fluoride in dimethyl 2-chloroethylene-l, 1-dicarboxylate by using the KF complex of dicyclohexano-18-crown-6 reaction (1).

$$ClHC = C(COOCH_3)_2 \rightarrow FHC = C(COOCH_3)_2 \quad ...(1)$$

In reaction (1), a crown provides solubility for an otherwise insoluble or marginally soluble salt. Use of crowns to transfer a salt from the solid phase into an organic phase is often referred to as solid-liquid phase-transfer catalysis. Such reactions may also be conducted by using liquid conditions. Generally, the most concentrated possible aqueous solution is contacted with an organic solvent containing the substrate. An example is the reaction between concentrated aqueous potassium cyanide and 1-bromooctane. The crown ether catalyst complexes the cation, transferring it from the aqueous to the organic phase. The cyanide anion accompanies the cation into the organic phase, where it is poorly solvated and therefore highly reactive. The cyanide ion and the organic bromide react to give 1- cyanooctane in high yield, reaction (2)

$$KCN + Br\text{–}C_8H_{17} \xrightarrow{\text{crown}} NC\text{–}C_8H_{17} + KBr \quad ...(2)$$

4. **Cation transport :** Because of the ability of crown ethers to complex cations and because of their obvious similarity to such marcocyclic ionophores as valinomycin, extensive studies of crowns as transport agents have been undertaken. Many such studies have been conducted in U-tube devices or variants therefore. In this experiment, a U-shaped tube is partly filled with an organic solvent like chloroform. The sidearms are then filled with water. A salt or mixture of salts is dissolved in one aqueous phase (source phase). The crown, cryptand or other ionophore is dissolved in the chloroform, which serves as a liquid membrane. The crown complexes the cation and conducts

it across the membrane to the other aqueous phase (receiving phase). Such experiments are used to team about the selectivities of ligands for cations and to understand the behaviour of membrane system. It is hoped that these membranes will be of commercial importance, perhaps in separating valuable metals from common ones.

5. **Indicators and Sensors :** Since crown ethers and related species complex, cations selectively, they can be used as sensors. Crowns have been incorporated into electrodes for this purpose, and crowns having various appended chromophores have been prepared. When a cation is bound within the macroring, a change in electron density is felt in the chromophore. The chromophores are often nitroaromatic residues and therefore highly coloured. The colour change which accompanies complexation can be easily detected and quantitated. It seems likely that such species will find future application in the rapid or automated analysis of various biological fluids.
6. **Toxicity :** Some crowns and cryptands like many other organic compounds, are irritants or toxic, 15-crown-5 has produced a slight redness in skin tests and is reported to be toxic in mice at a level of about O.Oloz/Ib (800 mg/kg) of body weight. This means that a 150-lb (70 kg) human would have to ingest about 2 oz (56 g) of this compound to be in danger. The toxicity of simple aspirin is about half this (that is, twice as much aspirin, 0.0267 oz/Ib or 1.75 g/kg, is required for the same toxic effect), so the crown toxicity levels are not generally high. Of course, not all crowns have been tested, and some may be dangerous.
7. **Analysis :** The receptor properties of the crown ethers and cryptands make these compounds potentially very useful in the area of analytical chemistry. They have been employed in extraction analysis, where the component to be determined is selectively extracted from the unknown mixture. Crown ethers have also been prepared possessing chromophores which change colour upon complexating, hence allowing easy detection of the particular metal ion. This technique has been used to measure the concentration of K^+ in blood serum. Exchange resins have been prepared bearing crown ethers and cryptands and these have been used to extract particular cations from aqueous solution. These resins can also be used in column chromatography where

the mixture of metal ions is separated by elution with water. Crown ethers and cryptands can also be used to advantage with ion selective electrodes.

Ion selective electrodes are very, convenient tools for the determination of ionic concentrations. These devices possess a .membrane containing a specific carrier compound. When the electrode is brought into contact with the ion (for which it is specific), the carrier renders the membrane permeable and an electrical potential is set up, proportional to the concentration of the ion present. The naturally occurring macrocyclic antibiotics (valinomycin and such) and the crown ethers shave been used as carriers in these electrodes. The synthetic compounds have the advantage that they can (in principle) be designed with particular characteristics. The ion-specific electrodes can be used to determine the stability constants of new crown ethers and cryptands with metal ions by measuring the free metal ion concentration in the presence of the macrocycle. Thus, well-characterized crown ethers and cryptand can be used to investigate the properties of newer members of the group.

8. **Isotope separation :** The discrimination of the crown ethers and cryptands between metal ions extends to an appreciable extent, toward the isotopes of a particular metal. When $^{48}Ca^{2+}$ and $^{40}Ca^{2+}$ are partitioned between water and chloroform in the presence of dicyclohexyl-18 crown-6, the organic layer is enriched in $^{40}Ca^{2+}$ by about 1 %. Thus, considerable enrichments could be achieved by successive liquid/liquid (countercurrent) extractions. Under similar conditions (2.2.1) cryptand (V) causes a 3% isotopic separation between $^{7}Li^{+}$ and $^{6}Li^{+}$.

9. **Medicine :** The ability of the crowns and cryptands to selectively complex metal ions suggests their use as drugs for the detoxification of organisms contaminated with radioactive or heavy metals. In has been shown that (2.2.2) cryptand facilitates the elimination of radioactive strontium- 85 and radium- 224 from rats and can also remove lead. This compound shows too many side effects to be used in the medical treatment of human beings, but illustrates a useful approach to these problems.

10. **Solubilization of Salt :** Complexation of the cations of metal salts by crown ethers of cryptands generally increases the solubility

of the salts in less polar solvents. Dramatic examples of this effect include the solubilization of potassium permanganate in benzene by dicyclohexyl-18-crown-6[0.02 M] and by (2.2.2) cryptands; (2.2.2) cryptand also causes a 10^4 increase in the solubility of barium sulphate in water, dissolving up to 50 g/litre.

11. **Anion Activation :** A very important consequence of the solubilization effect of the crown ethers and cryptands is that of anion activation. Complexation of the cation of a salt not only increases its solubility but also reduces ion pairing between the new larger cation and the anion. The poor solvation of anions in polar aprotic solvents gives rise to solutions of so-called naked anions which show a greatly enhanced nucleophilic reactivity. For example, the rate of reaction (4) is increased 280-fold by the addition of 5% dicyclohexyl- 18 crown-6. The crown ether acts in a catalytic role and can be recovered at the end of the reaction. On the laboratory scale such catalysis is of interest as it enables reactions to be carried out under milder conditions, perhaps thereby allowing the preservation of other functional groups in the reactant which would otherwise be destroyed. Also, the crown ethers have been shown to initiate novel nucleophilic substitution reactions. From the industrial view point such catalysis is of paramount importance, allowing reactions to be carried out with lower inputs of activation energy, hence conserving energy and reducing the need for capital-intensive high-temperature/ pressure reaction plants.

    In general, the cryptands are more effective than the metal crown ethers for producing naked anions because the metal cation is fully encapsulated, which reduces ion pairing even further. On the debit side, they are more difficult to synthesize and hence expensive [it should be pointed out that many crown ethers and cryptands, like (I) and (II), are commercially available compounds.] Metal complexes of crown ethers and cryptands are also effective lipophilic cations in the related field of phase transfer catalysis.

12. **Transport :** The transport of substrates through membranes is of fundamental importance in biology. The simplest mechanism of transport is that of facilitated diffusion, whereby transport of a substrate is assisted by a carrier molecule. A carrier must

possess some of the characteristics of a receptor, in particular showing a high specificity for the substrate. However, it is important that the affinity of the carrier for the substrate is not too high for if the complex is too stable there will be insufficient free carrier available to diffuse back across the membrane and complete the transport cycle. Also, the carrier and complex must have some solubility in the phase outside the membrane in order to facilitate complexation and decomplexation, and finally the rates of these processes must be rapid. In order to design a carrier to selectively transport one substrate from a mixture, it is necessary that the substrate is not only the most efficiently transported but also most strongly complexed.

MOLYBDENUM COMPOUNDS

Molybdenum forms compounds showing valencies of 2, 3, 4, 5 and 6. These compounds are mostly covalent, of which the hexavalent compounds are important. Molybdenum compounds of lower valencies are strong reducing agents.

Ammonium Molybdate, $(NH_4)_2MoO_4$

Preparation

Ammonium molybdate is obtained by dissolving molybdenum trioxide, MoO_3, in excess of hot concentrated ammonia solution. The solution on cooling deposits the crystals of normal ammonium molybdate.

$$MoO_3 + 2NH_4OH \rightarrow (NH_4)2MoO_4 + H_2O$$

The ordinary ammonium molybdate is the para molybdate, $(MH_4)Mo_7O_{20}.4H_2O$. This is obtained when MoO_3 is dissolved in dilute ammonia solution and solution evaporated to crystallization

$$7MoO3 + 6NH_4OH \rightarrow (NH_4)_6Mo_7O_{20} + 3H_2O$$

Properties

1. Ammonium molybdate is a white crystalline power; soluble in water but insoluble in alcohol.
2. On heating, ammonium molybdate is decomposed to MoO_3.

   $$(NH_4)6Mo_7M_{20}.4H_2O \rightarrow 7MoO_3 + 6NH_3 + 7H_2O$$
3. On heating ,at 500°C ammonium molybdate is first changed to reddish-brown dioxide, MoO_2 and at 1200°C to a grey powder of metallic molybdenum.

4. With phosphates in the presence of nitric acid, ammonium molybdate forms yellow precipitate of ammonium phosphomolybdate.

$$Ca_3\ (PO_4) + 6HNO_3 \rightarrow 3Ca(NO_3)_2 + 2H_3PO_4$$

$$H_3PO_4 + 12\ (NH_4)_2MoO_4 + 21HNO_3 \rightarrow (NH_4)_3PO_4.12MoO_3\uparrow$$
$$+ 21\ NH_4NO_3 + 12H_2O \qquad \text{Yellow ppt.}$$

5. With arsenic salt it forms yellow ammonium arsenomlobydate. Thus, when As_2S_5, As_2S_3 or is boiled with concentrated acid, arsenic acid, H_3AsO_4, is produced which gives yellow precipitate of ammonium arsenomolybdate with ammonium molybdate.

$$As_2S_5 + 10HNO_3 \rightarrow 2H_3AsO_4 + 10NO_2 + 2H_2O + 5S$$

$$H_3PO_4 + 12(NH4)_2MoO_4 \rightarrow (NH_4)_3\ AsO4.12MoO3 \downarrow$$
$$+ 21NH_4NO_3 + 12H_2O \qquad \text{Yellow ppt.}$$

6. On adding ammonium molybdate to a solution of stannous chloride, a blue coloured solution or precipitate of Mo_3O_8 is formed.

$$3(NH_4)_2 + MoO_4 + SnCl_2 + 8HCl \rightarrow Mo_3O_8 + 6NH_4Cl + 4H_2O + Sncl_4$$

Uses

(i) Ammonium molybdate is a very good analytical reagent. It is used in the detection and determination of phosphate and arsenic.

(ii) It is used as a catalyst for dehydrogenation and desulphurization in petroleum and coal industries.

Molybdenum Blue

Preparation

1. Treadwell in 1946 prepared molybdenum blue by the electrolytic reduction of ammonium paramolybdate in the presence of sulphuric acid.
2. When an acidified solution of a molybdate or molybdic acid is treated with reducing agents such as SO_2, H_2S, hydrazine, and glucose, an intense blue colouration is obtained due to the formation of mostly colloidal substance, known as molybdenum blue.

Properties

It is a blue powder soluble in water. The exact nature of this substance is not known. They vary in formula within the limits $MoO_{2.5}$ and MoO_3. Thus the mean oxidation state of the metal atom is between 5 and 6. It seems very probable though it is not completely proved that the aqueous solutions of the blue oxides are colloidal.

Molybdenum blue is amorphous towards *X*-rays. *Glemses* (1951) obtained some of its hydrates in crystalline forms such as $Mo_8O_{23}.8H_2O$, $Mo_4O_{11}H_2O$ and $Mo_2O_5(OH)_8$.

Bordeaux red, $Mo_5O_7(OH)_8$: When molybdic oxide, MoO_3 is reduced with $LiAlH_4$ an olive green compound of the composition $H_2Mo_5O_7(OH)_8$ is first obtained which then loses a molecule of hydrogen to form red coloured compound of the composition $Mo_2O_5(OH)_8$.

PER-OXYMOLYBDIC ACID AND PEROXY MOLYBDATES

When excess of hydrogen peroxide is added to acidified solutions of molybdate, a yellow or orange colour is obtained due to the formation of peroxymolybdates. These are derived from molybdates by the replacement of an oxygen atom by a peroxy group. The are generally very complex. The simplest peroxy molybdates are formulated as :

$$[MoO_4]^{2-} + 2H_2O_2 \rightleftharpoons [HMoO_6]^- + [OH]^- + H_2O$$

$$[MoO_4]^{2-} + 4H_2O_2 \rightleftharpoons [MoO_8]^{2-}\ 4H_2O$$

Peroxy molybdic acid $H_2MoO_5 2.5H_2O$ is an amorphous yellow powder.

POLY-ETHER COMPLEXES OF ALKALI AND ALKALINE EARTH METALS

Crown ethers are macrocyclic organic compounds generally composed of repeating ethylene (CH_2CH_2) units separated by noncarbon atoms such as oxygen, nitrogen, sulphur or phosphorus (I) and (II). Cryptands are similar to crown ethers but have a third organic chain attached at two points on the crown ring. Although, other carbon-containing subunits such as propylene ($CH_2CH_2CH_2$) or methylene (CH_2) may be included, these subunits are less common. The conformations of the ethylene unit are more favourable than those of propylene, and the ethylene unit is more stable to hydrolysis than is the O-CH_2-O linkage (an acetal). Organic units which are sterically equivalent to the ethylene unit, such as 1, 2 benzo, have been incorporated, as have a variety of carbohydrates (sugars) having adjacent hydroxyl groups. These substituents confer chirality on the resulting structure. Oxygen is incorporated most commonly into the ring, but numerous examples of compounds containing N, S, or P in place of O exist. The cryptands and crown ethers have the capability of forming stable and selective complexes with monoatomic or molecular ions and, as such, are potential receptor, carrier or catalyst molecules.

CLASSES OF MACROCYCLES

Several classes of macrocyclic polyethers are known. Substitution of oxygen by nitrogen, sulphur, or phosphorus leads respectively at azacrowns, thiacrowns or phosphacrowns. Likewise, when the ether link (–O–) is replaced by an ester link (–CO–O–), macrocyclic ester (lactone) analogs are formed. More than one ester unit and mixtures of esters or heteroatoms are common.

Cryptands : J.M. Lehn added a third strand to the simple marcocyclic polyethers and formed three-dimensional compounds based on the crown framework. Typically, two of the oxygen atoms across the ring from each other are replaced by nitrogen and a third ethyleneoxy chain is attached. Known as cryptands, these structures completely encapsulate cations and bind them strongly.

Systematic nomenclature of cryptands is complicated. Usually, a cryptand is presumed to be composed of ethyleneoxy chains attached to a nitrogen atom at either end. A typical cryptand would be $N[(CH_2CH_2)O_2CH_2CH_2]_3N$. Since each chain contains two oxygen heteroatoms, it is referred to as [2.2.2] crypland, or simply [2.2.2]. A cryptand composed of chains having two, three, and four ethyleneoxy units tied to nitrogen atoms at either end has one, two, and three oxygens respectively in the chains and is called (3.2.1) cryptand.

Larial Ethers : This name has been given to compounds like the cryptands, which have a ring and a third strand, but the additional chain is attached at only one point. This permits three-dimensional encapsulation of a cation while retaining a high degree of flexibility. A typical lariat ether. N–(2-methoxyethyl) monoaza 15-crown-5, and three typical cryptands are shown on the right side.

Spherands and Cavitands : Compounds other than simple cryptands have been developed which are capable of either partially (cavitands) or completely (spherands) enveloping cations. The donor atoms (O, N, S) are arranged so that they provide a solvadon sphere to the encapsulated cation.

The earliest example of a spherand, Lehn's all aliphatic spherand, is shown on right along with an all-aromatic spherand, also developed by D.J. Cram. Macrotricyclic compounds have properties similar to those of spherands. They consist of two crown ether-type structures bridged by two ethyleneoxy or diethyleneoxy strands. The structures are roughly tubeshaped.

Complexation Phenomena : It is the complexation of various species which makes this group of compounds interesting. A crown ether can be described as a doughnut which has an electron-rich and highly polar hole and a greasy or lipophilic exterior. As a result these compounds are usually quite soluble in organic solvent but accommodate positively charged species in their holes.

The complexation process shown in reaction (1) is usually characterized by a binding or stability constant (K_s) of the form k_c/k_d where k_c represents the rate constant for complex formation ($K_{complex}$) and k_d represents the rates of decomplexation ($k_{decomplex}$ or $k_{release}$).

The two important components of complexation are the magnitude of the equilibrium constant and the kinetics of the process. For simple crown ethers, k_c is usually very large and k_d is also large. These rate constants are solvent-dependent, but k_d decreases much more in nonpolar solvents then does k_c. Thus, k_d is smaller in polar solvents and larger in nonpolar solvents. Both k_c and k_d are lower for cryptands and spherands than for crowns, but k_d is lowered much more. As a result, binding constants (K_s) for the three- dimensional binders are generally much higher (they bind cations much more strongly) than for simple crowns.

Binding constants : Binding or stability constants can be assessed by two methods. One approach is to dissolve in water a cation such as Li^+ or Na^+ paired with a coloured organic anion like picrate. Shaking with an equal volume of organic solvent does not extract the salt. When a crown ether is added, the cation is complexed and transported into the organic phase.

The crown cation complex is accompanied by the coloured anion, so the extent of extraction can be determined by colorimetric analysis. The extraction constants are expressed numerically as the percent of total salt extracted. Such experiments suggest that crown ethers coordinate best with metallic cations whose diameters are similar in size to the crown hole. For example, 15-crown-5 binds Na^+ more strongly than K^+, and 1 S-crown-6 binds K^+ more strongly than either Na^+ or Li^+. The hole of 15-crown-5 Is estimated to be 0.20 nanometer across, and the cation diameter of Na^+ is about the same. The cation diameter of 8-Crown-6 is 0.27-0.30 nm. and the diameter of K^+ is near 0.27 nm. This hole-size relationship is observed generally when binding is assessed by the picrate extraction technique, and clearly obtains in general for the less flexible ligands like cryptands and spherands. It is not strictly obeyed by flexible crowns and lariat ethers, especially when binding is assessed by homogeneous binding-constant measurements.

Homogeneous binding constants are determined for a ligand and a cation present simultaneously in solution. These K_s values may by determined be a variety of techniques, but are most often obtained from either calorimetric or ion-selective electrode studies. In the series of simple crowns from 12-crown- 4 to 24 crown-8 homogeneous binding constants for K^+ are higher than for Na^+, Ca^{2+}, or ammonium, and 18-crown-6 is the best cation binder in the series. Homogeneous binding constants correlate well with extraction constants for less flexible ligands. It should also be noted that values for K_s, are highest (strongest association of ligand and cation) in the least polar solvents and lowest in solvents like water.

Complexing species : A variety of organic cations have been found to complex with ligands of this class. It has been suggested that for a host-guest interaction to occur, the host must have convergent binding sites and the guest must have divergent sites. This is illustrated by the interaction between optically active dibinaphtho-22-crown-6 and optically active phenethylammonimn chloride as shown below. The crown ether oxygen atoms converge to the centre of a hole and the ammonium hydrogens diverge from nitrogen. Three complementary O–H–N hydrogen bonds stabilize the complex. In this particular case, different steric interactions between the optically active crown and the enantiomers of the complex permit resolution of the salt.

Other organic cations have also been complexed, either by insertion of the charged function in the crown's polar hole or by less distinct interactions observed in the solid state. Arenediazonium cations, for example, are complexed in solution and in the solid-state by a various crown ethers. Also, a compound such as 18-crown-6 traps acetonitrile in its matrix when it crystallizes. Other molecules like nitromethane, thiourea, and dimethyl acethylenedicarboxylate have been trapped in crown crystals as well.

An area of particular research interest in recent years has been the construction of planar, macrocyclic ligands. These are special types of polydentate ligands in which the ligating atoms are constrained in a large ring encircling the metal atom. Examples are polyethers in which the ether oxygen atoms, separated by two methylene groups each, lie in a nearly planar arrangement about the central metal atom and the remainder of the molecule lies in a "crown" arrangement hence the name crownethers.

All of the oxygen atoms "point" inward toward the metal atom, and these macrocycles have the unusual property of forming stable complexes with alkali metals. This exceptional stability has been attributed to the close fitting of the alkali metal ion into the hole in the centre of the ligand. However some data seem to contradict this simple model. Although calculations indicate that the Li^+ should preferentially fit crown-4, in solution crown-4, crown-5, crown-6, and crown-7 all prefer K^+. Hancock has suggested that this may arise because the five membered ring formed when any of the above crown ethers binds to an alkali metal cation best fits the size of K^+.

Gas-phase studies show that crown-5 prefers Li^+ more than other alkali metal cations :

$$\text{Crown-5 : } Li^+ >> Na^+ > K^+ > Cs^+$$

On the other hand crown-6 shows more affinity for Na^+ and crown-7 prefers K^+:

$$\text{Crown-6 : } Na^+ \geq K^+ > Li^+ > Rb^+ > Cs^+$$

$$\text{Crown-7 : } K^+ > Na^+ > Rb^+ > Li^{++} > Cs^+$$

The difference in affinities shown in gas and solution phases suggests that solvent effects are quite important. These ligands have the unusual ability to promote the solubility of alkali salts in organic solvents as a result of the large hydrophobic organic ring. For example, alkali metals do not normally dissolve in ethers as they do in ammonia, but they will do so if crown ligands are present:

$$K + \text{crown} - 6 \xrightarrow[\text{or THF}]{Et_2O} [K(\text{crown} - 6)]^+ + [e(\text{solvent})]^- \qquad ...(1)$$

The ability to complex and stabilize alkali metal ions has been exploited several times to effect syntheses that might otherwise be difficult or impossible. Consider the compound Cs^+Au^-. Its existence might at first appear to be improbable since it is an ionic compound between two metals rather than between a metal and a nonmetal. Yet inspection of the ionization energy of cesium and the electron affinity of gold (or merely comparing the Pauling electronegativity of gold, 2.54, with that of iodine, 2.66) sparks curiosity:

It should be investigated. Indeed, based on the suggestion of the late Sir Ronald Nyholm, the first edition of this book (1972), as well as this edition, both mention the possibility of such a bond. The experimental stumbling block is obvious, however: If you mix two metals, how do you determine that you do indeed have an ionic compound and not an alloy? When cesium was allowed to react with gold, the conductivity in the melt was characteristic of an ionic compound. Further evidence was desirable, however. Two approaches were used. One was to take advantage of the solubility of alkali metals in liquid ammonia, the strong reducing power of the free electron and the stability of reduced species in this medium. (Note that Au^- should be a powerful reductant.)

The second was to stabilize metals, such as sodium, that might not otherwise react (recall that the ionization of any metal, even sodium or cesium, is *endothermic*) with macrocyclic polyethers to form $[Na(macrocycle)]^+ Au^-$ salts. A combination of these techniques with physical methods such as nonaqueous electrochemistry and those showing the anomalously low ESCA binding energy and the Mossbauer chemical shift were used to characterize the 1-gold. Unfortunately Sir Ronald did not live to see his prediction verified; he died in an automobile accident in 1971.

In a similar manner the so-called Zintl salts composed of alkali metal cations and clusters of metals as anions were known in liquid ammonia solution but proved to be impossible to isolate; Upon removal of the solvent they reverted to alloys. Stabilization of the cations by complexation with macrocyclic ligands allowed the isolation and determination of the structures of these compounds.

The general trait of crown ethers and cryptands (to be discussed later) to stabilize alkali metal salts has been extended to even more improbable compounds, the alkalides and electrides, which exist as complexed alkali cations and alkalide or electride anions. For example, we know that alkali metals dissolve in liquid ammonia (and some amines and ethers) to give solutions of alkali electrides :

$$M \xrightarrow{NH_3} M^+ + e^- \quad ...(2)$$

However, the situation is somewhat more complicated than Eq. (2) would indicate, because the electrons can react further with the metal to form alkalide ions :

$$M + e^- \rightarrow M^- \qquad ...(3)$$

Thus, in general, there are alkali metal cations, alkalide anions, and electride anions (and perhaps other minor species) in an equilibrium mixture dictated by various energetic factors. Note, for example, that because of the excellent solvating of the electride ion by liquid ammonia, the alkalide ions are not favoured in these solutions, and they are somewhat atypical. By using methylamine and ether solvents, even adding solvents of low polarity such as n-pentane, the equilibria can be shifted and crystals sometimes grown. The affinity of alkali metal ions for crown ethers and cryptands causes cationic complexes to form readily in solution. (Eq. (1) above). This, too will affect the equilibrium, with Eq. (2) being shifted to the right and therefore, Eq. (3) being shifted to the left, by the addition of these ligands. A rough generalization can be made that excess ligand will favour the formation of the electride inasmuch as all of the metal may be complexed as the cation.

$$\text{Cs + excess 18-6-crown-6} \rightarrow [\text{Cs (18-ctowa-6)}_2]^+ + e^- \qquad ...(4)$$

On the other hand, a mole ratio of 2:1, metal: ligand, tends to favour the alkalide :

$$\text{2Na + 18-crown-6} \rightarrow [\text{Na 18-(crown-6)}]^+ + \text{Na}^- \qquad ...(5)$$

Aside from the effects of ligand stoichiometry and the nature of the solvent, there are also differences in stability of the alkalide ions. The sodide anion is the most stable, and the ceside ion the least. Because of differential stabilities of the alkalide ions and differing stability constants of the [M (ligand)]$^+$ complexes, it is possible to form mixed metal compounds :

$$\text{Na + K 18-crown-6} \rightarrow [\text{K(18-crown-6)}]^+ + \text{Na}^+ \qquad ...(6)$$

As expected, all of these systems are strongly reducing and tend to decompose on exposure to air and moisture, so very careful work is necessary to study them. In a few cases, single crystals have been grown and structures determined crystallographically. Thus, crystalline [Na (macrocycle)]$^+$ Na$^-$ consists approximately of closed packed, large, complex cations with sodide anions in the octahedral holes. The alkalide ions can be found and measured, as expected they are considerably larger than the alkali metal cations.

In fact, the Cs^- anion is the largest known monoatomic ion with a radius of about 310-350 pm. In one salt there is anion-anion contact giving an effective radius of 319 pm. In addition to the usual difficulties of assigning an exact radius to an ion, the ceside anion has the added property that it is not only largest ion, but also the most polarizable.

The structure of $[Cs\,(18\text{-crown-}6)_2]^+\,e^-$ has been determined. Because the electride anions are extremely poor scatterers compared to the large cesium cation (and to a lesser extent the C and O atoms of the crown ether), the structure has the odd *appearance* of complexed metal cations with no corresponding anions. However, the most likely position of the electrons can be inferred from the presence of cavities of 240-pm radius; presumably the electrons are located in these cavities.

One final example of macrocyclic complexation will be given : the fluoride ion, F^-, should be a strong base and nucleophile. Normally, however, it does not show these expected properties because its very basicity attracts it to its countercation so strongly that it is ion-paired in solution and not free to react However, addition of, for example, crown-6 to a solution of potassium fluoride in benzene increases the solubility tenfold and also increases the nucleophilicity of the fluoride ion.

Earlier we saw that chelating ligands form complexes of greater stability than those of funidentate ligands. This greater stability was attributed primarily to entropy effects but enthalpy effects are of some importance. Macrocyclic ligands are even more stabklhan open-chain chelating ligands.

A thermodynamic comparison of 18-crown-6 complexes of Na^+, K^+ and Ba^{2+} with those of pentaglyme. $CH_3\ (OCH_2CH_2)_5\ OCH_3$. The additional stability is primarily an enthalpy effect due to preorganization of the macrocycle. This is not tc say that the conformation of 18-crown-6 is the same as that found in the complex.

Nevertheless, the energy required to rearrange the macrocyclic ligands for complex formation is less than the energy required to rearrange pentaglyme into a suitable conformation. In addition to their direct structural relationship to biological molecules, macrocycles such as the polyethers may provide clues to the discrimination shown by biological tissues toward various ions. This selectivity provides the so called sodium pump necessary for the proper Na^+/K^+ ionic balance responsible for electrical gradients and potentials in muscle action.

Table 2 : Thermodynamic contributions to the macrocyclic effect In complexes of 18-crown-6 and pentaglyme, $CH_3(OCH_2CH_2)_5OCH_3$, In methanol.

		Na^+	K^+	Ba^{2+}
log K_1	18-crown-6	4.36	6.06	7.04
	pentaglyme	1.44	2.1	2.3
	log K difference	2.92	3.96	4.74
ΔH	18-crown-6	–35.1	–56.0	–43.5
	pentaglyme	–16.7	–36.4	–23.8
	ΔH difference	–18.4	–19.6	–19.7
ΔS	18-crown-6	–33	–71	–13
	pentaglyme	–29	–84	–33
	ΔS difference	4	13	20

The ultimate in encirclement of metal ions by the ligand is shown by encapsulation reactions in which the ligand forms a three-dimensional cage about the metal. The resulting cage called a *clathro-chelate* or a *cryptate*. One class of cryptate-forming ligands of the type $N(CH_2CH_2OCH_2CH_2OCH_2CH_2)_3N$ has been called "football ligands" because the polyether bridges between the two nitrogen atoms resemble the seams of a football.

Ligands of this type form exceptionally stable complexes with alkali metals and show high selectivity when the size of the "football" is adjusted to fit the desired cation. The effectiveness of THF and dimethylethers of ethylene- and diethyleneglycols as media for reactions involving sodium metal may be due in part to sought solubility of Na but the driving force is undoubtably the complexation of Na^+ by the chelate ethers.

The most important ethers are the crowns and cryptates (O, N) which bind M^+ strongly and often with high selectivity. The affinity of such a ligand for an ion is strongly dependent on how well the ion fits into the cavity that the ligand can provide for it At the same time the strength of complexation and to some extent the selectivity also depend on the solvent.

Illustrations of selectivity, provided by cryptate-221 and-222. For cryptate-222, for example, K^+ fits the cavity very well, but Li^+ and Na^+ are too small to make good contacts with the oxygen atoms and Rb^+ and Cs^+ and too large to enter without appreciable steric strain.

WALFRAM (TUNGSTEN) ELEMENTS

The element and its ores were described by Agaricola in 1556, as a form of wolf (it eats up tin as wolf eats up the sheep). The German name Wolfram and the symbol (*W*) are derived from this notion. Scheelite in 1781 demonstrated that scheelite contained calcium and a peculiar acid which he called tungstic acid. He obtained the yellow acidic trioxide WO_3 from the ore Scheelite.

Tungsten is not found in native state but is present in the combined state. Important minerals of this element are:

1. Wolframite	$FeWO_4$ and $Mn\ WO_4$
2. Scheelite	$CaWO_4$
3. Hubnerite.	$MnWO_4$
4. Tungsite or Wolfram ochre	WO_3

Extraction

The extraction of tugsten from its ores involves the following steps.

1. **Crushing :** The It is crushed to a fine powder.
2. **Concentration :** It is concentrated by electromagnetic separators as the ore contains magnetic impurities. The powdered or is dropped over a travelling belt moving round electromagnetic rollers. The magnetic constituents of the ore are attracted and collected in a heap nearer to the magnets while non-magnetic constituents drop always in another separate heap.
3. **Roasting :** The crushed ore is roasted with soluble carbonate. Soluble sodium tungstate is formed. Iron and manganese get oxidised to their respective oxides.

$$4\ FeWO_3 + 2Na_2CO_3 + O_2 \rightarrow 4Na_2WO_4 + 2Fe_2O_3 + 4CO_2\uparrow$$

$$2MnWO_4 + 2Na_2CO_4 + O_2 \rightarrow 2Na_2WO_4 + MnO_2 + MnO_2 + 3CO_2\uparrow$$

4. **Extraction with hot water :** The roasted mass is then extracted with hot water. Sodium tungstate being soluble, dissolves. Oxides of iron and manganese are removed by filtration.
5. **Treatment with acid :** The filtrate containing sodium tungstate is reacted with acid. Tungsten trioxide (WO_3) is precipitated.

From Scheelite

(i) When Scheelite is fused with potassium fluoride, the formation of potassium tungstate and calcium fluoride takes place. CaF_2 is insoluble where as k_2WO_4 is soluble in water. The fused mass is thus extracted with water. The solution thus obtained on acidification yields the precipitate of tungsten trioxide.

$$CaWO_4 + K_2F_2 \rightarrow K_2WO_4 + CaF_2$$

$$K_2WO_4 + 2HCl \rightarrow 2KCl + WO_3 + H_2O$$

(ii) Then scheelite is digested with concentrated hydrochloric acid, Most of the impurities go into solution while the residue consists of tungstic acid and silica.

Tungstic acid gets separated by dissolving it in ammonium hydroxide with the formation of ammonium tungstate. The solution on crystallization yields crystals of ammonium para tungstate. These crystals on heating at 600°C yield tungsten trioxide.

$$CaWO_4 + 2HCl \rightarrow H_2WO_4 + H_2O + CaCl_2$$

$$7H_2\ WO_4 + 6NH_4OH \rightarrow (NH_4)_6W_7O_{24} + 10H_2O$$

$$(NH_4)_6W_7O_{24} \rightarrow 6NH_3 + 7WO_3 + 3H_2O$$

From Wolframite

(i) When the concentrated or is mixed excess of sodium carbonate and the mixture is fused in reverberatory furnace in presence of excess of air, iron and manganese present in the ore get converted into ferric oxide (Fe_2O_3) and manganese dioxide (MnO_2) respectively. Both of these oxides are insoluble in water.

$$4FeWO_3 + 4Na_2CO_3 + O_2 \rightarrow 4Na_2WO_4 + WO_4 + 2Fe_2O_3 = 4CO_2 -$$

$$2MnWO_4 + 2Na_2CO_3 + O_2 \rightarrow 4Na_2WO_4 + 2MnO_2 + 4CO_2 -$$

Sodium tungstate is soluble in water. The fused mass is extracted with water when sodium tungstate passes into solution. The solution is acidified with hydrochloric acid and boiled to yield tungstic acid.

$$Na_2WO_4 + 2HCl \rightarrow 2NaCl + H_2WO_4$$

The tungstic acid on strong heating forms tungsten trioxide.

$$H_2WO_4 \rightarrow H_2O + WO_3$$

The oxide thus obtained is made to dissolve in ammonium hydroxide. The solution is filtered to remove insoluble (silica etc.) material. The solution on acidification yields the precipitate of hydrated trioxide. By this repeated precipitation a purer sample of WO_3 is obtained.

(ii) From wolframit, sodium tungstate is obtained by the application of the above method by fusing the ore with Na_2CO_3. The solution having sodium tungstate is treated with calcium chloride to precipitate calcium tungstate. The precipitate is then decomposed with HCl or HNO_3 to yield tungsten trioxide.

$$Na_2WO_4 + CaCl_2 \rightarrow CaWO_4 + 2NaCl$$

$$CaWO_4 + 2HCl \rightarrow H_2WO_4 + CaCl_2$$

$$\downarrow$$

$$H_2 + WO_3$$

Wolframite.

Fused with Na_2CO_3 in reverberatory furnace in presence of excess of air. Fused mass extracted with water.

Metallurgy : Tugsten may be obtained in metallic state by several methods:

1. **Reduction of :** WO_3 or H_2WO_4 with (a) hydrogen, (b) carbon, (c) Zinc, (d) aluminium, (e) magnesium, and (f) silicon,

   $$WO_3 + 3H_2 \xrightarrow{1250.C} W + 3H_2O$$

2. **Reduction of chloride :** WCl_6 with H_2 or Na

   $$WCl_6 3H_2 \rightarrow W + 6HCl$$

   $$WCl_6 + 6Na \rightarrow W + 6NaCl$$

3. **Alumino-thermic process :** When a mixture of WO_3 mixed with aluminium powder is ignited by magnesium cartridges, W is set free.

   $$WO_3 + 2Al \xrightarrow{450\text{-}600C} W + Al_2O_3$$

4. **Electrolytic method :** Electrolysis of WO_3 dissolved in boric acid is carried out at 1200° – 1400°C.

The properties of tungsten vary with the degree of purity and the method of production.

1. Pure tungsten is a steel grey metal.
2. It has a bright metallic lustre.
3. Its melting point is the highest of known metals (m.p. = 3370°C and b. P. = 4830°C).
4. Tungsten has high tensile strength and is a dense metal.
5. The specific gravity of metal is 19.30 approximately equal to that of gold.
6. In massive form it is white but a very hard powder of grey to brownish black colour can be prepared.
7. In powder from the metal is hard, brittle and crystalline and in compact or ductile form it is much softest and tougher.

Chemical

1. **Action of air or oxygen :** Tungsten remains unaffected by air or oxygen at ordinary temperature, but at higher temperatures it oxidises rapidly :

$$2W + 3O_2 \rightarrow 2WO_3$$

2. **Action of water :** The red hot tungsten filament decomposes water vapour :

$$W + 2H_2O \rightarrow WO_2 + 2H_2$$

3. **Action of halogen :** Tungsten is not affected by halogens at ordinary temperature but when heated to redness the halogens attack it.

   (a) **Fluorine :** It attacks tungsten at an ordinary temperature with incandescence forming volatile fluoride.

   (b) **Chlorine :** Chlorine attacks the metal forming hexachloride, $WCl_{5.}$

$$W + 3Cl_2 \xrightarrow{250.C} WCl_6$$

   (c) **Bromine :** At an elevated temperature the metal is attacked by bromine forming WBr_5.

$$2W + 5Br_2 \rightarrow 2WBr_5$$

   (d) **Iodine :** The metal ombines with iodine at red heat forming $Wl_{2.}$

4. **Action of non-metals :** Molten sulphur and phosphorus attack the metal slowly but the vapours react vigorously.

5. **Action of halogen acids**

   (i) **HF :** The metal is not affected by hot or cold hydrofluoric acid.

   (ii) **HCl :** This acid has no action on the metal up to 500 – 600°C.

6. **Action of oxy-acids :** Sulphuric acid and nitric acid have little or no action except when boiling. Hot and concentrated acids have slow action forming WO_3.

7. **Action of mixed acids :** Pure tungsten is resistant to the action of most single acids but mixture of HNO_3 and HF dissolves it readily. Chromic acid makes the metal passive.

8. **Action of alkalies :** The powdered metal dissolves in fused nitrates, peroxides, alkaline carbonates, hydroxide and acid sulphate but alkaline solution has no effect.

$$W + 2KOH + 2H_2O \rightarrow K_2WO_4 + 3H_2$$

9. **Action of oxidising agents :** Oxidising agents like hydrogen peroxide has no action on the metal but it is readily attacked by heating with (i) NaO_2 and (ii) PbO_2, forming WO_3.

Uses

Tungsten finds several uses. These are :

1. **In filaments for electric lamps :** Because of very high melting point and very little volatility at high temperature, tungsten is used for making filaments of electric lamps.

2. **For alloying steel (Tungsten steel) :** Steel Containing 15 to 20% of tungsten and 5% chromium are very hard and retains its temper at high temperatures. Tungsten steel is employed for cutting tools.

3. **Thermocouples: Tugsten :** Molybdenum thermocouples are used for high temperature work.

4. Oxides of tungsten are used in ceramics to impart different shades in glass and porcelain.

5. Calcium tungstate is used for the preparation of X-ray fluorescent screens.
6. Lead-tungsten alloy finds use in making bullets and shots.

   Tungsten electrodes prove very efficient in electrochemical processes where resistance to corrosion is needed.
7. Tugsten has been tried as a catalyst for numerous reactions.
8. Tungsten carbide is used as a cutting agent and abrasive, as it is the hardest substance known.
9. Sodium tungstate is used (i) for decolorisation of acetic acid, (ii) for the water proof cloth, and (iii) as mordant for silk.

TUGSTEN BRONZES

The tungsten bronzes are non–stoichiometric compounds of general formula M_nWO_3 ($0 < n \leq 1$). The actual range of composition of the tungsten bronzes is approximately $Na_{0.3}WO_3$ to $Na_{0.95}WO_3$. The tungsten bronzes are obtained by reducing sodium tungstate with hydrogen at red heat. These substances are chemically inert and have a bronze like appearance. The colours vary greatly with composition form golden-yellow to blue violet. Tungsten bronzes with $n > 0.3$ are extremely inert and have semi-metallic properties such as metallic lustre and good electrical conductivity in which in charge carriers are electrons. Those tungsten bronzes with $n < 0.3$ are semi-conductors, insoluble in water and all acids except hydrofluoric acid. They can be oxidised by oxygen in presence of base to give tungstate (IV).

$$4NaWO_3 + O_2 + 4NaOH \rightarrow 4Na_2WO_4 + 2H_2O$$

Tungsten bronzes have been put to many uses, especially in making acid resistant material and as semi-conductors.

TUNGSTEN COMPOUNDS

Tungsten shows velencies of 2, 3, 5 and 6 in its compounds. The compounds of tungsten resemble closely molybdenum compounds and being most stable. Compounds of lower valencies are reducing agents.

Tungsten Trioxide WO_3

Preparation

1. By the hydration of tungstic acid.

$$H_2WO_4 \rightarrow WO_3 + H_2O$$

2. By roasting tungsten sulphide in a current of air

$$2WS_2 + 9O_2 \rightarrow 2WO_3 + 4SO_3$$

3. By heating ammonium tungstate or ammonium oxy flurotungstate, the yellow powder of WO_3 is obtained

$$(NH_4) WO_4 \rightarrow WO_3 + 2NH_3 + H_2O.$$

Properties

1. It is a bright yellow powder which assumes orange colour when hot.
2. It melts to green liquid at 1472°V and boils at 1750°C.
3. It is insoluble in water but its hydrated form is soluble forming tungstic-acid.

$$WO_3 + H_2O \rightarrow H_2WO4$$

4. It is not affected by acids except HF.

$$WO_3 + 2H_2F_2 \rightarrow \underset{\text{Hydrofluotungsticacid}}{H_2WO_2F_4} + H_2O.$$

5. Hydrogen reduces the trioxide successively to blue oxide, a brown oxide and then the metal.

   On heating with $CaCl_2$ in absence of air, it forms tungstyl chloride.

$$CaCl_2 + 2WO_3 \rightarrow CaWO_4 + WO_2Cl_2$$

6. When heated with H_2S or CS_2, it forms sulphide.

$$WO_3 + 3H_2S \rightarrow WS_3 + 3H_2O$$

Uses

1. As mordant in dyeing and ceramic industry.
2. As pigment in oil and water colours.

Tungstic acid H_2WO_4

Preparation

1. Bo treating sodium or ammonium tungstate with hot sulphuric acid.

$$Na_2WO_4 + H_2SO_4 \rightarrow H_2\ WO_4 + Na_2SO_4$$

It form tungstates with alkalies.

$$Ha_2WO_4 + 2N_aOH \rightarrow H_2O + WO_3$$

$$H_2WO_4 + 2NH_4OH \rightarrow (NH_4)_2WO_4 + 2H_2O$$

$$H_2WO_4 + 2NH_4OH \rightarrow (NH_4)_2WO_4 + 2H_2O$$

Composition

The compositions of various tungstic acids are:
Meta tungstic acid ... $WO_3.H_2O$ or H_2WO_4
Para tungstic acid ... $WO_3.2H_2O$ or H_4WO_5
Ortho tungstic acid ... $WO_3.3H_2O$ or H_6WO_6

Uses

1. In ceramic industries.
2. In making pigments.

White tungstic acid : When solutions of paratugstates or normal tungstates are treated with sufficient excess of acids in cold, a white voluminous precipitate of white tungstic acid is formed. It has been found to be a 'gel' of tungsten trioxide. On warming with acids, this precipitate changes into yellow tungstic acid. White tungstic acid is fairly soluble in water and has acidic reactions. It easily passes into colloidal state but precipitation occurs on increasing the hydrogen ion concentration by addition of an acid.

Meta tungstic acid : It prepared by the action of H_2S on lead meta tungstate or by the action of H_2SO_4 on barium meta tungstate. After filtering off the precipitate, the aqueous solution is evaporated yielding large rhombohedral crystals of metatungstic acid. Its composition corresponds to the formula $4WO_3$. $9H_2O$. Meta tungstic acid is considered to be a heteropoly acid as it is isomorphous with tungsto phosphoric acid $H_3[P(P(W_3O_{10})_4]$.$24H_2O$. This acid may be named as tetra (tri-tungsto) dihydrogen acid.

Tungsten blues : These substances are blue in colour and of variable composition depending upon the mode of their preparation. They contain tungsten in an intermediate oxidation sate of 5 of 6. These blue compounds re generally produced by mild reduction of WO_3 or tungstic acid in solution or in contact with water. It has been seen that WO_3 *becomes* blue precipitate is stable in air. It is soon decolorised and change into yellow tungstic acid H_2WO_4. The stabler form of the tungsten blue has also been obtained by the action of lithium aluminium hydride or atomic hydrogen on WO_3. The unstable form of tungsten blue is nearer in composition to tungstic acid.

Trivalent tungsten : When the solution of potassium tungstate in HCl is reduced with tin, dark green powder of the composition $K_2[W(OH)Cl_5]$ is first formed, when reduction is carried further, the yellow green salt $k_2[W_2Cl_9]$ is obtained. It crystallises out as green yellow hexagonal plates.

Malleable tungsten : Reduction of tungstic oxide yields the metal in the form of a very fine powder. It is made compact and ductile as follows; The metal powder is pressed into bars and sintered at high temperature. The hot bars are annealed and hammered. The process of heating annealing and hammering is repeated till the rods become fairly ductile and can be drawn into wires. By repeatedly drawing of wires, the metal becomes very flexible and has a tensile strength.

Sodium Tungstate, Na_2WO_4

Preparation

The tungsten trioxide WO_3, dissolves in aqueous sodium hydroxide from which the simple sodium tungstate can be crystallised by evaporating the solution to crystallization point.

$$WO_3 + 2NaOH \rightarrow Na_2WO_4 + H_2O$$

However, when WO_3 is heated with sodium carbonate solution sodium paratungstate, $Na_{10}W_{12}O_{41}28H_2O$ is formed. When sodium paratungstate solution is boiled with WO_3 sodium metatungstate, $Na_2W_4O_{13}$. $10H_2O$ is obtained.

Properties

(i) Sodium tungstate is white crystaline powder. It is water.

(ii) When a solution of sodium tungstate is made weakly acid, polymeric anion solution tungstic acid in formed. At room temperature yellow tungstic acid, $WO_3.2H_2O$ (or $H_2WO_4.H_2O$) crystallize, but from hot solutions $WO_3.H_2WO_4$ is obtained.

(iii) When a solution of sodium tungstate and sodium phosphate is acidified with sulphuric acid phosphotungstic acid, $P_2O_5.24WO_3.63H_2O$, is formed.

Uses

Sodium tungstate is used as a mordant and in rendering-flannelette non-inflammable.

4

Some Elements of Titanium Group

INTRODUCTION

The Gergor in 1789 made the discovery of titanium from magnetic sand-Menachanite. Klaproth in 1793, while analysing the mineral Rutile, discovered the new metal Hunter in 1910 prepared titanium metal of 99.9% purity by reducing $TiCl_4$ with metallic sodium.

Occurrence

Titanium is commonly regarded a rare element but it is one on the nine most abundant elements.

Important Minerals

1. Oxide Ores : (a) Rutile TiO_2
   (b) Brokite TiO_2
   (c) Anatase TiO_2
2. Titanate Ores : (a) Ilmenite $FeTiO_3$
   (b) Pervoskite $CaTiO_3$
3. Silicate Ores : (a) Sphene $CaTiSiO_6$
   (b) Guarinite $CaTiSiO_5$.

Extraction

There are many methods for extracting titanium from its ores. The ores generally used are :

(i) Rutile, and

(ii) Ilmenite.

Rutile is a stable substance and causes difficulty in complete decomposition by chemical treatments. Ilemnite may be decomposed with greater ease; but due to large quantity of iron present, it creates difficulties in purification.

1. Wohler's process

The finely powdered ore is fused with thrice its weight of K_2CO_3. The fused mass is cooled, powdered and digested with dilute HF and evaporated so that potassium fluotitanate is formed.

$$\underset{\text{(Rutile)}}{TiO_2.(Fe_2O_3)} + K_2CO_3 \rightarrow K_2TiO_3 + Fe_2O_3$$

$$\underset{\text{(Ilmentie)}}{2FeTiO_3} + 2K_2CO_3 \rightarrow 2K_2TiO_3 + Fe_2O_3 + CO_2 + CO$$

$$K_2TiO_3 + 6HF \rightarrow K_2TiF_6 + 3H_2O$$

$$K_2TiF_6 + 4NH_4OH \rightarrow TiO_2 + 2H_2O + 2KF + 4NH_4F$$

The residue is boiled with water and filtered hot so that K_2TiF_6 goes into solution. When the filtrate is cooled, then crystals of K_2TiF_6 separate out.

Ilmenite Fe TiO_3

|

Fused with potassium carbonate and digested with hydrofluoric acid soln.

|

$K_2TiF_6 + Fe_2O_3$

$[FeTiO_3 + K_2CO_3 + H_2F_2 \rightarrow Fe_2O_3 + K_2TiF_6 + CO_2\uparrow + H_2O]$

↓ Extracted with boiling water

↓ Residue Fe_2O_3 Crystals

↓ Boiling solution K_2TiF_6

$K_2TiF_6 \xleftarrow{\text{Cooling}} K_2TiF_6$

| + NH_4OH

↓

↓ ppt. Hydrated TiO_2

↓ Soln. NH_4F,KF

↓ ignition

TiO_2

$$K_2TiF_6 + 6NH_4OH \rightarrow 2KF + NH_4OH + TiO_2 + 2H_2O,$$

K_2TiF_6 crystals are dissolved in hot water, acidified with dil. HF. The solution is concentrated, and cooled when pure crystals of K_2TiF_6 are obtained. K_2TiF_6 crystals are again dissolved in boiling water. To this ad NH_4OH in order to precipitate $Ti(OH)_4$ or hydrated TiO_2. The ppt is washed, dried and ignited to get TiO_2.

2. Rose's Process

Finely powdered mineral is heated to redness in a stream of H_2S so that iron sulphide is precipitated. The solid mass is cooled and heated with dil. HCl and filtered. The insoluble residue is TiO_2 while iron passes into solution as $FeCl_2$. TiO_2 is filtered, washed and dried.

$$\underset{\text{(Ilmenite)}}{FeTiO_3} + H_2S \rightarrow FeS + TiO_2H_2O$$

$$\underset{\text{(Rutile)}}{TiO_2(Fe_2O_3)} + 3H_2S \rightarrow TiO_2\ 2FeS + S + 3H_2O$$

$$FeS + 2HCl \rightarrow FeCl_2 + H_2S$$

MODERN METHODS

1. Fusion with Na_2O_2

The powdered ore is fused with Na_2O_2 when sodium titanate is formed. The fused mass is dissolved in water and filtered. The filtrate having sodium titanate is then boiled so that it suffers hydrolysis and hydrated TiO_2 is precipitated. It is separated, dried and ignited to get pure anhydrous TiO_2.

$$2TiO_2(Fe_2O_3) + 2Na_2O_2 \rightarrow 2Na_2TiO_3 + O_2 + Fe_2O_3$$

$$FeTiO_3 + Na_2O_2 \rightarrow Na_2TiO_3 + FeO + O$$

$$Na_2TiO_3 + H_2O \rightarrow TiO_2 + 2NaOH$$

2. Fusion with $KHSO_4$

The powdered ore is mixed with six times its weight or $KHSO_4$. The fused mass is boiled with water and filtered, $(NH_4)_2S$ solution is added to the filtrate so that $Ti(OH)_4$ and iron sulphide are precipitated. The precipitates are separated by filtration. The ppt. is dissoved in HCl whereby iron sulphide dissolves but TiO_2 remains undissolved.

$$FeTiO_3 + 2KHSO_4 \rightarrow K_2TiO_3 + FeSO_4 + H_2SO_4$$

$$FeSO_4 + (NH_4)_2S \rightarrow FeS + (NH_4)_2SO_4$$

$$K_2TiO_3 + H_2O + (NH_4)_2S \rightarrow K_2S + 2NH_4OH + TiO_2$$

$$FeS + 2HCl \rightarrow FeCl_2 + H_2S$$

Metallurgy

The preparation of pure titanium is difficult on account of its high melting point and its tendency to form compounds with oxygen, nitrogen, carbon monoxide, etc. It also readily forms alloys with metals.

The following methods are used for the preparation of titanium from its different compounds.

GERMANIUM COMPOUNDS

Germanium like tin and lead yields two series of compounds, germanium (Ge^{+2}) and germanic (Ge^{+4}) the latter is more stable the former. As its salts are easily oxidized, they act as strong reducing agents.

Oxides of germanium : Germanous oxide, or germanium monoxide, GeO can be obtained by heating germanic oxide, GeO_2 with magnesium. It is a dark grey powder and readily soluble in hydrochloric acid to give germanous chloride. It reduces the solutions of chromates, permagnates and gold salts.

Germanic oxide or germanium dioxide : GeO_2 is prepared by heating the metal in oxygen or by roasting the sulphide in air. It is also prepared by heating the metal with nitric acid or by the action of water on the tetrachloride. It occurs a as dense white powder, which is very stable. It is an acidic oxide.

Halides of germanium—Fluorides : Germanous fluoride, GeF_2 is obtained if potassium fluogermnate is warmed in a stream of hydrogen.

Germanium tetrafluoride, or germanic fluoride : GeF_4 is prepared by dissolving the dioxide in hydrofluoric acid. It is a hygroscopic substance.

Chlorides—Germanous chloride : $GeCl_2$ can be obtained by the action of hydrogen chloride on germanous sulphide or by dissolving germanous oxide in HCl acid.

Germanium tetrachloride or germanic chloride : $GeCl_4$ is prepare by dissolving the dioxide in HCl acid or by heating germanium in a stream of chlorine. It occurs as a colourless mobile liquid, fuming strongly in air. It gets decomposed by water yielding the germanium dioxide. It is soluble in alcohol, benzene, chloroform, ether and CS_2. With conc. HCl is yield chlorogermanic acid, $H_2[GeCl_6]$.

Sulphides : Germanous sulphide. GeS and germanic sulphide, GeS_2 get precipitated by hydrogen sulphide from the corresponding salts. These occur as white solids and like tin sulphides they dissolve in yellow amm. Sulphide, yielding thiogermanates.

MATERIAL IS ITS DIOXIDE

1. Reduction with Carbon

Titanium can be obtained by reducing titanium dioxide with carbon at 2500°C in an electric furnace. The titanium thus obtained is largely contaminated with the oxide and carbide.

$$TiO_2 + 2C \rightarrow Ti + 2CO$$

2. Gold Schmidt's Process

Titanium dioxide is made to mix with aluminium and the mixture is ignited with a suitable fuse. Along with aluminium, barium peroxide is generally added to facilitate the reduction.

$$3TiO_2 + 4Al \rightarrow 2Al_2O_3 + 3Ti$$

$$TiO_2 + 2Al + BaO_2 \rightarrow Ti + Al_2O_3 + BaO$$

The titanium thus obtained is alloyed with 5% of Al.

COMPLEX COMPOUND

1. BerZelius Process

Berzelius in 1825 obtained titanium by the potassium fluotitanate K_2TiF_6, with sodium.

$$K_2TiF_6 + 4Na \rightarrow 2KF + 4NaF + Ti$$

2. Hunter's Process

Hunter in 1916 succeeded in preparing pure titanium by heating the tetrachloride with metallic sodium in a steel bomb.

$$TiCl_4 + 4Na \rightarrow Ti + 4NaCl$$

3. Billy's Process

In 1921, Billy prepared the metal by the reduction of titanium tetrachloride with sodium hydride.

$$TiCl_4 + 4NaH \rightarrow Ti + 4NaCl + 2H_2$$

4. Van Arkel and DeBoer Process

In 1924 these scientists prepared pure crystalline titanium by thermal decomposition of sodium tetraiodide (TiI_4) by the filament growth process. It involve heating the iodide in vacuum and decomposing it on an incandescent tungsten filament. The method is based on the fact that the decomposition temperature of iodide is very much lower than the volatilization temperature of the elementary titanium. Hence it is deposited on the filament in crystalline form

$$\underset{(\text{Impure})}{Ti + 2_2} \xrightarrow{150-250^oC} TiI_4 \xrightarrow{1100^oC} \underset{(\text{pure})}{Ti + 2I_2}$$

5. W. F. Kroll's Process

The most important industrial method of obtaining titanium is by reduction of tetrachloride by molten magnesium or Sodium in an atmosphere of argon. Kroll in 1937 prepared titanium by this technique. The resulting metal is spongy, but it can be melted electrically in a copper crucible, the walls of which are cooled by water so as to produce a lining of solidified titanium. It can then the cast into ingots.

$$TiCl_4 + 2Mg \rightarrow 2MgCl_2 + Ti$$

6. High grade titanium can also be obtained by the electrolysis of potassium fluorotitanate K_2TiF_6 dissolved in molten sodium chloride. The electrolysis is carried out in graphite crucible, which acts as anode, in an atmosphere of argon, using steel or graphite cathode.

Preparation of Fused Metal :The fused titanium in the metallic form is obtained by melting the powdered material, in a vacuum are furnace. Alternatively, the powder may be dressed into bars at 70,0000 atmospheric pressure and used as electrodes in the sides of are. The metal fuses in the form of globules on the sides of electrodes.

None Refining for the Preparation of Pure Titanium : This method uses a unique process (zone refining) to melt a bar of metal while the metal acts a its own crucible, thus preventing contamination by any containing vessel. By this method super pure titanium metal is obtained. The success of the process depends upon the fact that most impurities in metal have a preference for either solid or liquid state of the metal. Iron, a common impurity in titanium, has preference for the liquid state of titanium. When a bar of impure titanium is melted progressively from end to end, the iron concentrates in the liquid titanium and follows the molten zone to the end of the bar. This end is then cut off and discarded leaving the rest of the titanium bar in pure condition.

Uses

1. As Alloys

(a) Alloys of titanium with tantalum and niobium resist corrosion and possess unusual physical properties.

(b) Ferrotitanium is used in steel industry and scavanger, a cleaning agent and a deoxidising agent in the metallurgical operation.

(c) Titanium alloys may be employed as filaments in vacuum tubes. This replaces costly platinumiridium alloy.

2. As Metal

(a) Titanium may be used as a substitute for other elements like molybdenum, tungsten or tantalum in certain high melting point alloys.

(b) a titanium cathode is used in a special photoelectric cell.

(c) Extremely pure titanium is used in making incandescent filaments.

3. Titanium Soaps

are found useful in preventing setting of paint in water proofing, in, stabilizing suspensions and in imparting special properties to paints and varnishes.

4. Smoke Screen

Titanium terachloride is used in war times for the production of irritating smoke clouds and dense fog formation.

5. Mordants and Dyes

Titanium compounds find a variety of application of dyeing cotton, wool, paper, leather and rubber, TiO_2 is largely used in opacifying and colouring in ceramics and glass industries. Postassium titanium oxalate finds extensive applications both for dyeing and mordanting of leather. It gives various shades from yellow to brown.

6. Paints and Pigments

Titanium dioxide (Titanox) is extensively used in the manufacture of paints and pigments.

7. Other Uses

(i) Titanium carbide is used as a substitute hard alloys for machine tools which is a recent development.

(ii) Titanous salts are used in volumetric analysis as a reductant

(iii) TiO_2 finds uses in cosmetics as whitener in face powder, ceramics, etc.

PHYSICAL PROPERTIES

(1) Metallic titanium has silvery lustre and when polished resembles to steel appearance. Amorphous form is a dark grey powder resembling iron (reduced) in appearance.

(2) It is brittle at ordinary temperature but at low red heat it is readily malleable. The pure metal is very ductile.

(3) Titanium melts at 1670°C and boils at a temperature above 3275°C.

(4) It is a light metal of specific gravity 3.5.

(5) The fused metal is very hard and crystalline and scratches quartz.

(6) It is paramagnetic and sometimes shows a yellow tint resembling bronze.

(7) It shows pyrophoric properties, when rubbed against steel, *i.e.*, gives bright sparks

(8) Specific heat of titanium varies with temperature. It rises rapidly with increase in temperature.

CHEMICAL PROPERTIES

(1) Reaction with Oxygen

Titanium is stable in air and there is no reaction of oxygen up to 120°C in oxygen. The metal catches fire and burns with brilliant light forming TiO_2.

$$Ti + O_2 \rightarrow TiO_2$$

(2) Reaction with Nitrogen

At 800°C it burns in nitrogen forming titanium nitride TiN. In air it takes up both nitrogen and oxygen which can not be expelled by vacuum treatment. In air it forms both oxide and nitride.

$$2Ti + N_2 \rightarrow 2TiN$$

(3) Reaction with Hydrogen

Titanium adsorbs hydrogen with the formation of an indefinite compound, made up of 1 part titanium and 1.73 part of H_2.

(4) Reaction with Water

Titanium does not decompose water at ordinary temperature but decomposes steam.

$$2Ti + 3H_2SO_4 \rightarrow Ti_2(SO_4)_3 + 3H_2$$

(b) Dil. hydrofluoric acid :

$$2Ti + 6HF \rightarrow 2TiF_3 + 3H_2$$

(c) Dil. nitric acid :

$$2Ti + 6HNO_3 \rightarrow 2Ti(NO_3)_3 + 3H_2$$

(d) Dil acetic acid.

$$2Ti + 6CH_3COOH \rightarrow 2(CH_3COO)_3Ti + 3H_2$$

Hot and concentrated HCl evolves hydrogen.

$$2Ti + 6HCl \rightarrow 2TiCl_3 + 3H_2$$

(6) Reaction with Halogens

Titanium combines with halogens especially chlorine forming $TiCl_4$.

$$Ti + 2Cl_2 \rightarrow TiCl_4$$

(7) Reaction with Boron, Carbon and Silicon

Titanium forms hard refractory :

(1) Boride TiB

(2) Carbide TiC

(3) Silicide TiSi.

Compounds of Titanium

Titanium forms compounds in the oxidation states of +2, +3 and +4 but the compounds containing tetravalent titanium are usual and stable under many conditions. The compounds of trivalent titanium are generally obtained with difficulty by the reduction of Ti(IV) compounds and are rather unstable in aqueous solution.

Titanium (IV) has a strong tendency to form complex anions such as titnate ion $[TiO_4]^{3-}$, flouorotitanate ion $[TiF_6]^{2-}$, chlorotitanate $[TiO_6]^{2-}$, and sulphato titanate on $[Ti(SO_4)_3]^{2-}$ all of which are trivalent anions.

The soluble salts of Ti(IV) have a strong tendency to get hydrolysed in aqueous solutions and form the precipitate of TiO_2 or hydrate dioxide but when hydrolysis is incomplete titanyl salts of the type $TiOX_2$ containing the trivalent titanyl caution $[TiO]^{2+}$ are produced.

Salts of Ti(III) also show some tendency to form complexes which are unstable. Salts of trivalent titanium possess reducing properties. The values of normal oxidation potentials for various changes in the valence states of titanium are as given below :

(i) $Ti \rightarrow Ti^{2+} + 2e^-$ $E^o = + 1.75$ volt

(ii) $Ti^{2+} \rightarrow Ti^{3+} + e^-$ $E^o = + 0.37$ volt

(iii) $Ti^{3+} \rightarrow Ti^{4+} + e^-$ $E^o = + 0.20$ volt

(iv) $Ti + 2H_2O \rightarrow TiO_2 + 4H^+ + 4e^-$ $E^o = + 0.25$ volt

(v) $Ti^{3+} + H_2O \rightarrow [TiO]^{3+} + 2H^+ + e^- E^o = + 0.10$ volt

Tetravalent titanium compounds are well known and are important. trivalent titanium compounds which are commonly known as titanous compounds are important reductants.

(TITANOX) TIO_2

Preparation

1. This can be prepared from rutile (TiO_2).

Flow Sheet

Finely powdered rutile $\xrightarrow[Na_2CO_3]{\text{fused with}}$ Fused mass

| dissolved in
↓ HF

Soln. in HF

| Extracted with

TiO_2 (ppt.) $\xleftarrow[NH_4OH]{\text{Added}}$ K_2TiF_6 Soluble

Filtrate (Rejected)

2. By heating ilmenite ($FeTiO_3$) ore in a stream of chlorine and HCl gas.

$$2FeTIO_3 + 4HCl + Cl_2 \rightarrow 2FeCl_3 + 3TiO_2 + 2H_2O$$

3. By decomposing titanium tetrachloride with the help of steam at red heat.

Properties

1. It is an amorphous white powder, yellow when hot.
2. Crystalline forms of TiO_2 occur n nature.
3. Its specific gravity lies between 3.88-4.24 depending on temperature.
4. It is insoluble in water but dissolves in acids with difficulty.

$$\underset{(hot)}{TiO_2 + 2H_2SO_4} \rightarrow \underset{\text{Titanyl sulphate}}{TiO(SO_4)}$$

5. It reacts with dry chlorine to form chloride.

$$TiO_2 + 2Cl_2 \rightarrow TiCl_4 + O_2$$

6. It can be reduced by carbon at very high temperatures.

$$TiO_2 + C \rightarrow Ti + CO_2 \uparrow$$

Uses

1. It is extensively used as pigment in paints and varnishes.
2. In the manufacture of white leather.
3. In glass industry
4. In rubber industry to soften rubber.
5. It is largely used in cosmetics as a whitener.

TITANIUM TETRACHLORIDE, $TICL_4$

Preparation

1. It is prepared by passing chlorine over heated titanium or mixture of titanium dioxide and carbon:

$$Ti + 2Cl_2 \rightarrow TiCl_4$$

$$TiO_2 + 2C + 2Cl_2 \rightarrow TiCl_4 + 2CO$$

2. It is also prepared by passing vapours of chlorine or carbon tetrachloride at high temperature over TiO_2.

$$CCl_4 + TiO_4 \rightarrow TiCl_4 + CO_2$$

Properties

1. It is colourless, strongly fuming liquid.
2. It has melting point of - 33°C and boling pint of 16.4°C.
3. It is soluble in water but hydrolysed in excess to form TiO_2
4. It reacts with ammonia to form smoke cloud. Reaction with ammonium chloride and hydrochloric acid gives yellow crystals of $(NH_4)_2TiCl_6\ 2H_2O$.

Uses .

1. This is used with ammonia to form smoke clouds.
2. It is chiefly used as the starting material for the production of metallic titanium.

COMPOUNDS OF TITANIUM (III)

In general, Ti(III) compounds have been found to be more ionic than Ti(IV) compounds. Titanium (III) ion, Ti^{3+} exists in aqueous solution as the blue violet $[Ti(H_2O)_6]^{3+}$ ion.

Due to their strong tendency to change to Ti(IV) compounds. Ti(III) compounds exhibit strong reducing properties.

$$Ti_2O_3 + O \rightarrow 2TiO_2$$

Titanium (III) chloride, $TiCl_3$

Anhydrous $TiCl_3$ may be prepared by passing vapours of titanium (IV) chloride mixed with excess of hydrogen through a red hot but heated to about 647°C.

$$TiCl_4 + 1/2H_2 \rightarrow TiCl_3 + HCl$$

Anhydrous $TiCl_3$ is a violet power.

Hydrated titanium (III) chloride exists in two forms-violet coloured $[Ti(H_2O)_6]^{3+}(Cl^-)_3$ and green coloured $[Ti(H2O)_5]^{2+}(Cl^-)_2$. Both the forms disportionate on heating to $TiCl_4$ and $TiCl_2$.

Titanium (III) Oxide. Ti_2O_3

The anhydrous oxide may be prepared by heating titanium (IV) oxide in a stream of hydrogen to 1027°C.

$$TiO_2 + H_2 \rightarrow Ti_2O_3 + H_2O$$

It is a highly stable oxide attacked only by nitric acid.

Titanium (III) Hydroxide, $Ti(OH)_3$

It is basic in nature and is a stronger base that $Ti(OH)_4$.

Titanium (III) Complexes : Ti(III)

Is known to form a number of complexes both cationic as well a anionic. An important example of the cationic complexes is the hexahydrate, $[Ti(H_2O)_6]^{3+}$. It is associated with chloride ions in complex salts of formula $[Ti(H_2O)_6]Cl_3$.

Two well known examples of anionic complexes are the ions hexafluorotitanate (III) $[TiF_6]^{3-}$, and pentachloroaquotitanate (III), $[TiCl_5(H_2O)]^{2-}$. Both of them are associated with alkali metal cations in crystalline state.

Compounds of Titanium (II)

The + 2 oxidation state of titanium is not very stable. Two important compounds formed by Ti(II) are given as follows :

(1) Titanium (II oxide, TiO)

It is a basic oxide and is ionic in character. It is a non-stoichiometric compound with the composition TiO_{075}. As it gets readily oxidized to TiO_2, it acts as a strong reducing agent. For example, it reduces water.

$$TiO + H_2O \rightarrow TiO_2 + H_2$$

Titanium (II) chloride, $TiCl_2$. Due to its strong tendency to get oxidized to TiCl4, it exhibits as a strong reducing agent. It is an ionic compound.

It undergoes disproportionation to form $TiCl_4$ and Ti.

$$2TiCl_2 \rightarrow TiCl_4 + Ti.$$

Detection

(i) Titanium salts give a white precipitate of phosphate with sodium phosphate which is insoluble in acetic acid.

(ii) Titanium salts give orange red precipitate with hydrogen peroxide.

(iii) Titanium salts give brown precipitate with postassium ferrocyanide in a slightly acidic medium.

(iv) Titanium salts give an intense blood red colour with thymol or salicylic acid in concentrated sulphuric acid.

Spot Tests

(i) *Hydrogen peroxide test :* A drop of the hydrochloric or sulphuric acid test solution is treated with a drop of 20% hydrogen peroxide on a spot plate, an intense yellow colour forms (M. Fujimoto, 1957).

(ii) *Morin test :* One drop of the test solution in hydrochloric acid, is treated on filter paper with one drop of morin in methyl alcohol, a yellow-brown fleck develops if titanium is present. (Murata and Yamauchi, 1963).

(iii) *Acetyl acetone test :* One drop of the test solution in 2N–HCl is mixed with a drop of acetyl acetone in a micro test tube or on a spotting plate, a red colour indicates the presence of Ti(III) salts. (Linden and Boef, 1967).

Estimation

(i) *Gravimetrically :* Titanium may be estimated by adding KIO_3 to a solution of titanium sulphate containing a small quantity of nitric acid. The precipitate, $Ti(IO_3)_4 . 3KIO_3$ is filtered, washed and dissolved in HCl to which sulphurous acid is added. Finally the titanium is precipitated with NH_4OH washed, dried, ignited and weighed as TiO_2.

(ii) *Volumetrically :* It can be determined by reducing the compounds to titanous solution and then titrating with postassium permanganate solution.

ZIRCONIUM DIOXIDE, ZRO_2

It is also known as zirconia. It occurs in nature as Beddeleyite from which it can be obtained in a pure state as described earlier.

ZrO_2 can also be obtained when the metal is heated in air or oxygen and also by the action of heat on the hydride, carbide, nitride, nitrate or hydroxide of zirconium.

It is very hard, infusible white powder which is insoluble in water. It forms zirconates when fused with alkali hydroxides and carbonated. When heated, it becomes incandescent at high temperatures.

It is used as a valuable refractory material for furnace lining. It makes a good paint.

Hydroxides

Zirconium seems unable to form a true hydroxide because two or more hydroxyl groups in the presence of each other on the same zirconium atom are unstable.

Zirconium hydroxide $Zr(OH)_4$

Preparation

It is precipitated in cold by adding ammonium hydroxide to a solution of zirconium tetrachloride.

$$ZrCl_4 + 4NH_4OH \rightarrow Zr(OH)_4 + 4NH_4Cl$$

Properties

It is white gelatinous precipitate amphoteric in behaviour. It forms salts with acids and alkalies. When $Zr(OH)_4$ is heated upto 200°C, it forms metal zirconic acid $ZrO(OH)_2$ known also as zirconyl hydroxide. But on strong heating metal zirconic acid yields ZrO.

$$ZrO(OH)_2 \rightarrow ZrO + 2H_2O$$

It is used for purification of sewage.

Halides : Zirconium forms various halides of the type ZrX_4, ZrX_3, and ZrX_2 where X stands for the halogen atom.

Zirconium Tetrafluoride, ZrF_4

Preparation

It is prepared dissolving zirconia (ZrO_2) inHF.

It can also be prepared by the action of anhydrous HF on $ZrCl_4$.

$$ZrCl_4 + 4HF \rightarrow ZrF_4 + HCl$$

Properties

It is a white crystalline substance which sublimes on heating. It forms a hydrate, $ZrF_4.3H_2O$. With liquid ammonia it forms a white powder of $5ZrF_4.2NH_3$.

ZrF_4 absorbs HF in cold forming hydrofluozirconic acid, H_2ZrF_6.

$$ZrF_4 + 2HF \rightarrow H_2ZrF_6$$

Zirconium Tetrachloride, $ZrCl_4$

Preparation

It is prepared by passing chlorine or hydrogen chloride over red hot zirconium.

$$Zr + 2Cl_2 \rightarrow ZrCl_4$$

It can also be prepared by passing a steam of chlorine and carbon monoxide over ZrO_2 at 400°C.

$$ZrO_2 + 2Cl_2 + 2CO \rightarrow ZrCl_4 + 2CO_2$$

Properties

It is a white crystalline solid which fumes in air. In water it gets hydrolysed. It is soluble in alcohol and ether.

$$ZrCl_4 + H_2O \rightarrow ZrOCl_2 + 2HCl$$

It absorbs ammonia forming $ZrCl_4$, $2NH_3$; $ZrCl_4$. $4NH_3$ $ZrCl_4$. $8NH_3$ etc. It sublimes at 330°C. When heated with reducing agents like magnesium or aluminium it gives trichloride which on heating forms di and tetra chlorides.

$$2ZrCl_3 \rightarrow ZrCl_2 + ZrCl_4$$

It is used as a catalyst in the Friedel Craft's reaction.

Compounds of Lower Valence State

Zirconium tetrachloride, $ZrCl_3$ and dichloride $ZrCl_2$: When zirconium is heated in vapours of zirconium tetrachloride, the dichloride $ZrCl_2$ is formed as a black solid. When zirconium tetra chloride is heated with aluminium powder and aluminium chloride in a sealed tube to 350°C, both $ZrCl_2$ and dark red brown trichloride $ZrCl_3$ are formed. Zirconium trichloride is also formed by careful heating of the tetrachloride with zirconium metal.

Zirconium Tri-bromide

$ZrBr_3$ is formed by passing a mixture of zirconium tetra bromide vapours and hydrogen over an aluminium wire heated to 450°C. Zirconium dibromide is also formed along $ZrBr_3$.

Zirconium Tri-iodide

ZrI_3 and zirconium diodide ZrI_4 are also produced together by heating zirconiumtetra iodide with zorconium metal. In 1952, Ehrlich reported the formation of the trifluoride ZrF_3 which decomposed to elements on being sublimed in vacuum.

HAFNIUM

Discovery

In 1845, Savanberg declared that zircon was made up of two elements zirconium and the other named norium. In 1852, Sorby from spectroscopid studies also confirmed that zircon was composed of two elements and called the new element as Jargonium. In 1911, Urbain by spectrochemical and other methods of analysis established the presence of a new element inzirconium ores. In 1923, Coaster and Heavers, analysed zirconium minerals and showed that many of them contained upto 30 percent of the new metal. They named the new element hafnium from Hafnia, which is an ancient Latin name Copenhagen and also found is atomic number to be 72.

Occurrence

It occurs inzirconium minerals. Actually it was first discovered in these minerals by X-ray method. In the earth's crust, it occurs to the extend of 0.002 per cent. Its richest minerals are aviate which is zirconium, thorium ortho-silicate (Zr, Th, Hf) SiO_4, malacone, nageite, etc.

Separation of Zirconium and Hafnium

Zircon contains about 0.5 – 2.0% hafnium and all extraction methods for zirconium also lead to the extraction of hafnium. As the chemistry of these two metals is so similar that no direct chemical method can be employed for their separation and independent extraction. The only way lies in their slight difference in extraction behaviour or in their solubility of their compounds. Different methods given so far are :

1. Fractional Crystallisation of the Double Fluorides

It is possible to get reasonable separation by repeated crystallisation of the double fluorides from water. The solubilities of $(NH_4)_2$ $[Zr_6]$ and $(NH_4)_2$ $[HfF_6]$ in water are given below (in moles/litre):

$(NH_4)_2ZrF_6$	0.611	at 0°C
$(NH_4)_2HfF_6$	0.891	at 0°C.

In practice the mineral is fused with potassium hydrofluoride and hexafluozirconate and hexafluohafanate are separated from each other due to difference in their solubility. Fractional crystallisation method is now-a-days replaced by ion-exchange and solvent exchange methods.

2. Extraction of Zirconyl Chloride with Theiocynate-rich Methyl Isobuty Ketone (MIBK)

Crude zirconyl chloride ($ZrOCl_2$) is treated with counter current stream of thiocyanate rich methyl isobutyl ketone. Hafnyl chloride goes into the MIBK. Hafnium is preferentially extracted being contaminated with less than 2% zirconium. MIBK extract is then treated with aq HCl and then with aq H_2SO_4. By this treatment hafnium is brought into the aqueous layer. It is then precipitated with NH_4OH which on ignition yields oxide. The inexactitude zirconium is recovered as ZrO_2.

3. Elution of Tetra Chlorides from Silica Gel Column

Anhydrous methanol solution of the tetrachloride of Zr and Hf is passed through a silica gel column. The fixed chlorides are then washed first with 2N.HCl in methanol which removes $ZrCl_4$. Finally elution is done with $7N.H_2SO_4$ which removes $HfCl_4$.

4. Ion-exchange Method

The best method for separating zirconium and hafnium is based on the application of ion-exchange method. In this method, a mixture of $(NH_4)ZrF_6$ and $(NH_4)_2HfF_6$ is passed through anion-exchange resin of the type $[RNMe_3]Cl$ when the hexafluoroanions of zirconium and hafnium are absorbed by the resin but at a different capacity. Then the adsorbed anions are eluted with an aqueous mixture of HF and HCl which releases the hexafluoroanions from the resin. The anions, so formed, are eluted one after the other.

$$2[RNMe_3]Cl + (NH_4)_2ZrF_6 \rightarrow [RNMe_3]\ ZrF_6 + 2NH_4Cl$$

$$2[RNMe_3]Cl + (NH_4)_2HfF_6 \rightarrow [RNMe_3]\ HfF_6 + 2NH_4Cl$$

$$[RNMe_3]ZrF_6 + 2HCl \rightarrow H_2ZrF_6 + 2[RNMe_3]Cl$$

$$[RNMe_3]HfF_6 + 2HCl \rightarrow H_2HfF_6 + 2[RNMe_3]Cl.$$

FRACTIONAL PRECIPITATION METHODS

(i) *De Boer's method :* Hafnium can be separated from zirconium by the action of water on a sulphuric acid solution of the phosphates of the two elements. On dilution hafnium concentrates in the first fraction.

(ii) *Method of Caster and Haversy :* Zirconium phosphate $ZrO(H_2PO_4)_2$ and hafnium phosphate. $HfO(H_2PO_4)_2$ are dissolved in hydrofluoric acid when the complex phosphofluozirconates and hafnates are formed. To this mixture an equivalent amount of boric acid or some other boron compound is added, when a phosphate gets precipitated which is richer in hafnium than the solution from which it was originally precipitated. By this procedure a concentration of 60 per cent hafnium can be obtained.

Separation as Oxalates

Hafnium and zirconium is separated in the form of the double oxalates. The mixed salts are precipitated with ammonia, and then dissolved in oxalic acid. Thus a mixture of ammonium zirconium oxalate and ammonium hafnium oxalate is obtained. This mixture when crystallized, hafnium accumulates in the mother liquor.

Fractional Volatilization of Halides Van Arkel and De Boer Method

Hafnium is also separated from zirconium by taking advantage of high volatility of the hafnium double chloride. It is found that the hafnium tetrachloride with phosphorous pentachloride ($2HfCl_4 . PCl_5$) is more volatile than the corresponding zirconium compounds ($2ZrCl_4.PCl_5$) and hence distills over earlier. Similarly, the double fluorides can be used for separation by fractional volatilization.

Ion Exchange Methods

Ion exchange process is successfully used for the separation of zirconium and hafnium, *i.e.,* by elution of tetra positive ions by the cation exchange resin Dowex-50, with hydrochloric acid solution yields good results.

Metallurgy

(1) *Reduction of the oxide :* Hafnium oxide is reduced with sodium, calcium or magnesium. Oxide of hafnium is made to heat strongly with any one of the metals mentioned, when hafnium is obtained.

$$HfO_2 + 4Na \rightarrow 2Na_2O + Hf$$

$$HfO_2 + 2Ca \rightarrow 2CaO + Hf$$

$$HfO_2 + 2Mg \rightarrow 2MgO + Hf$$

(2) *Reduction of the tetrachloride :* Hafnium is also obtained by reducing the tetrachloride, $HfCl_4$ by sodium.

$$HfCl_4 + 4Na \rightarrow 4NaCl + Hf$$

Or by Van "Arkel and de Boer method *i.e.,* by passing the vapour of the tetraiodide, HfI_4 over a heated tungsten filament, when decomposition of iodide occurs and hafnium gets deposited on the filament.

$$HfI_4 \rightarrow Hf + 2I_2$$

Properties

Like zirconium, hafnium is silvery white, hard, high-melting metal (m.p. 2200°C and b.p 5000°C). Its chemical properties have been found to be similar to that of zirconium.

Compounds

Like zirconium, hafnium forms the compounds such as HfO_2, HfX_4 HfC, etc. Their preparations and properties are quite similar to the corresponding zirconium compounds.

Uses : It is mainly used in atomic power engineering for absorbing neutrons. It is also used in electronics as cathodes of television tubes.

VANADIUM, NIOBIUM AND TANTALUM

General Study of Vanadium Group

This group consists of transition elements, vanadium niobium and tantalum. They all possess the outer configurations d^3s^2.

Element	*Atomic No.*	*Electronic configuration*	*Ionisation potential (eV)*	*Electronegativity*
Vanadium	23	[Ar] $3d^3\ 4s^2$	6, 7, 14.6, 29.3, 48, 65.2	1.6
Niobium	41	[Kr] $3d^3\ 4s^2$	6.8 14.3, 25.0, 38.3, 50	1.6
Tantalum	73	[Xe] $4f^{14}\ 5d^3s^2$	7.8, 16.2	1.6

General feature of transition elements are observed in this groups also. On the whole the differences between the first two members are rather wide but the differences between the second and the third are minimal. All the three elements exhibit several valences.

Vanadium	+5	+4	+3	+2	+1	0	–1
Niobium	+6	+4	. .	. .	+1	. .	–1
Tantalum	+5	+4	. .	. .	+1	. .	–1

The highest oxidation state is acidic and the lower ones becomes progressively more basic. The lower oxidation states are naturally reducing the character. The lower valences, however, become less stable with increasing atomic number. It should be seen that the +3 state in these elements is caused by the inertness of the S^2 electron pair of the s^2d^3 outer configuration. The overall ionisation potentials for withdrawal of all five electrons are too much high to be compensated by lattice energies. Therefore, following Fajans generalisation all pentavalent compounds are covalent. With decrease in valence, the element becomes more basic and ionic behaviour predominates. In extremely low oxidation state, covalence predominates.

Atomic radii and ionic radii increases from vanadium to niobium but thereafter it is steady in tantalum. This steady size but a huge increase in atomic weight from 93 (niobium) to 181 (tantalum) leads to a two-fold increase in density. The standard potentials are negative which indicate same reducing ability of metals but the elements easily become passive in the presence of acids.

Element	*Atomic Radius (Å)*	*Ionic (M^{5+}) Radius*	*M.P. (°C)*	*B.P. (°C)*	*Density (gm/ml)*	*E° volts* $M^{5+} + 5e \rightleftarrows M$
Vanadium	1.36	0.56	1919	3000	6.1	–
Niobium	1.46	0.70	2468	4927	8.4	–0.6
Tantalum	1.46	0.73	2996	6300	16.6	– 0.7

CHARACTERISTICS OF TRANSITION ELEMENTS

Many of the compounds of the elements are coloured. Thus vanadium is colourless in +5 state ($3d^0$) blue or green in +4 state ($3d^1$) green in +3 state ($3d^2$) and violet in +2 state ($3d^3$) etc.

Furthermore, magnetic properties of their compounds are indicative of zero, one, two and three unpaired spin in the oxidation states +5, +4, +3 and +2 respectively. Once again it should be remembered that spin pair will favoured in niobium and tantalum than in vanadium. For example,

VCl_4 is paramagnetic but $NbCl_4$ is diamagnetic through poylmerisation and spin-spin interaction.

VANADIUM

In 1801 Del Rio of Mexico announced the discovery of new element in certain lead ores from Zimpan, Mexico. He suggested the name Erythronium meaning red. Its salts turned red when heated with acid. In 1830 Sefstrom found new metal in swedish iron one, which he proposed to call "Vanadium" after the Scandinavian goddess of love and beauty, "Vanadis". In the same year Wohler found that both erythronium and vanadium were infact identical. In 1987, Roscoe gave it the position in the periodic table along with phosphorus and arsenic in the V group.

Occurrence

Vanadium is not found in free state. Its compounds are widely distributed in the earth's crust. Important minerals are :

1. Vanadinite – $3Pb_3 (VO_4)_2 PbCl_2$ or $Pb_5 (VO_4)_3 Cl$
2. Carnotite – $K_2O. 2UO_3. V_2O_5. 3H_2O$
3. Patronite – V_2S_5
4. Descloizite – $4(Pb,Zn)O.V_2O_5. H_2O$
5. Asphalite – V_2O_5

Extraction

(1) *From Patronite :* The ore is first concentrated and roasted. The roasted ore is fused with a mixture of sodium carbonate and nitrate, when sodium vanadate is formed. This is extracted with boiling water and to the solution, ammonium chloride is added when an orange coloured precipitate of ammonium metal vanadate (NH_4VO_3) is obtained. This on heating yields vanadium pentoxide.

$$2V_2S_5 + 15O_2 \rightarrow 2V_2O_5 + 10SO_2$$

$$V_2O_5 + 2Na_2CO_3 \rightarrow 2Na_3VO_4 + 3CO_2 \uparrow$$

$$Na_3VO_4 + NH_4Cl + H_2O \rightarrow NH_4VO_3 + NaCl + 2NaOH.$$

$$2NH_4VO_3 \rightarrow V_2O_5 + 2NH_3 + H_2O.$$

This is then converted into ferro-vanadium by reduction with iron oxide and carbon, in an electric furnace using carbon electrodes. Some lime or fluorspar is added as a flux.

(2) *From Carnotite :* The ore is leached with conc. HCO_3 and heated at 100°C. The solution after filtration is neutralised with sodium hydroxide. It is then treated with barium chloride and sulphuric acid to precipitate radium and barium sulphates. The clear liquid is decanted off and treated with an excess of boiling solution of sodium carbonate so as to precipitate iron, calcium and most of the aluminium. The filtrate is then neutralised with nitric acid and treated with sodium hydroxide so as to precipitate uranium as sodium uranate. It is filtered and filtrate is treated with nitric acid and ferrous sulphate, which precipitates vanadium as iron vanadate.

Carnotite

↓ Treat with HCO_3 at 100°C

↓	↓
Residue	*Solution*
	Neutralize with NaOH and treat with $BaCl_2$ and H_2SO_4
↓	↓
Ppt	*Filtrate*
$BaSO_4$ and $RaSO_4$	UO_2SO_4 and $VOSO_4$
	Treat with Na_2CO_3 in excess
↓	↓
Ppt	*Filtrate*
Fe, Ca and Al	Sodium uranyl carbonate

$[Na_4UO_2(CO_3)_3]$ and sodium vanadate (Na_3VO_4)

	Neutralize with HNO_3 and treat with NaOH
↓	↓
Ppt:	*Filtrate:*
	Treat with HNO_3 and $FeSO_4$
↓	↓
Ppt:	*Filtrate:*
Iron vanadate	(Reject)

(3) *From Vanadinite :* Johnston's Process. Vanadinite ore is digested with HNO_4 and filtered H_2S is then passed into the filtrate so that Pb, as etc. are precipitated. Theses precipitates are filtered off; the filtrate

is evaporated to dryness. The dry residue is then boiled with a solution of ammonium carbonate when ammonium vanadate $(NH_4)_3VO_4$ separates out as crystals. These crystals on ignition yield V_2O_5.

Preparation of Ferro-Vanadium

The crude vanadate (*e.g.,* iron vanadate) or the pentoxide is mixed with iron ore or scale, lime or fluorspar and coke and heated in an electric furnace. Alternatively, reduction may be carried out by the Goldschmidt's aluminothermic process.

$$\begin{cases} V_2O_5 + 5C \rightarrow 2V + 5CO \uparrow \\ Fe_2O_3 + 3C \rightarrow 2Fe + 3CO \uparrow \end{cases}$$

$$3V_2O_5 + 10Al \rightarrow 6V + 5Al_2O_3$$

$$Fe_2O_3 + 2Al \rightarrow 2Fe + Al_2O_3$$

Metallurgy

Preparation of pure metallic vanadium has presented numerous difficulties.

(1) Marden and Rich Method

Marden and Rich reduces vanadium pentoxide with a mixture of $CaCl_2$ and calcium in a sealed bomb. A little sodium or potassium was added to remove last trace of air and moisture.

$$V_2O_5 + 5Ca \rightarrow 3V + 5CaO.$$

The bomb was heated to 900°C for an hour, cooled and charge dug out.

(2) Jantesh Method

It consists in reducing ferro-vanadium in a current of chlorine, when vanadium tetra chloride is formed as a dark red liquid. This is distilled to get-rid off from ferric chloride and then reduced by heating the vapour with hydrogen above 600°C.

$$2VCl_4 + 4H_2 \rightarrow 2V + 8HCl$$

This is most satisfactory method at present for the preparation of metal.

(3) De Boer and Van Arkel's Process

100% pure vanadium is obtained by this method. In this completely pure vanadium is deposit over an electrically heated vanadium wire mounted in an evacuated quartz bulb when vapours of VI_2 come in contact with the glowing wire, VI_2 is thus thermally decomposed into vanadium and iodine and malleable vanadium is deposited on the filament.

$$VI_2 \rightarrow V + I_2$$

(4) Bread and Crook's Method (1954)

It consists in reducing the vanadium pentoxide with metallic calcium in an inert atmosphere of helium or argon. A small quantity of iodine is added to the charge which supplies extra energy for the combination.

$$V_2O_5 + 5Ca \xrightarrow{I_2} 2V + 5CaO$$

PROPERTIES

Physical

It is a greyish white infusible metal. It is a 'hard' but light metal. It is a good conductor of electricity but allows the current to flow in one direction. It is paramagnetic.

Chemical

It is unreactive at room temperature, but combines with other elements readily on heating. Thus, on heating in air it is oxidised to form the pentoxide, in chlorine to VCl_4 while with nitrogen it forms VN. It also combines with hydrogen, phosphorus, arsenic, bromine and silicon. With carbon it forms a stable carbide, V_4C_3 It does not dissolve in clod mineral acids but dissolves in perchloric acid and in ammonium persulphate to form metavanadic acid, HVO_3.

It dissolves slowly in hot concentrated sulphuric acid, hot hydrochloric acid and hot dilute nitric acid. The soultions contain the pervanadyl ion, VO^+_2. It dissolves in hot fused alkalis to form corresponding vanadates and liberating hydrogen.

$$2V + 6NaOH + 2H_2O \rightarrow 2Na_3VO_4 + 5H_2$$

It is not affected by steam.

Uses

(1) Vanadium pentoxide is used as a catalyst in oxidation reaction. For example, in the manufacture of sulphuric acid by the contact process.

(2) Ferro vanadium is used as scavenger in the steel industry and the addition of small amount of vanadium 01-0.3 percent gives though close-gained steel with high tensile strength which is used in the construction of motor parts.

(3) Vanadium carbode is used in the manufacturer of incandesscent lamps.

(4) Vanadates have also been used in photography as developer and sensitizers. they have also been employed as driers for linseed oil paints.

(5) Vanadium salts are poisonous when taken in large amounts. Organo-vanadium compounds are used for specific diseases in small doses. They can also be employed as antiseptic and a skin stimulant. In view of the toxicity of vanadium compounds, they may possibly be used as fungicides and insecticides.

(6) Vanadium alloyed wit cobalt and iron has been recently reported (1964). The alloy possesses the highest residual mangnetism and at the same time is so ductile and malleable that it can be rolled to the thickness of a hair for use in red switches.

Compounds of Vanadium

Vanadium is notable for the number of oxidation states which it exhibits. Thus oxides and halides corresponding the oxidation states of 2, 3, 4 and 5 are known. The compounds in which vanadium exhibits lower oxidation states (2, 3, and 4), are usually powerful reducing agents.

Oxides of Vanadium

The oxides known are V_2O_2, V_2O_3, V_2O_4 and V_2O_5. A semi-oxide V_2O is also known.

Hypovanadous oxide, VO or VO or V_2O_2. It is obtained as a black powder by the reduction of higher oxides with potassium. Heating V_2O_3 with vanadium at low pressure also forms VO.

$$V_2O_5 + 6K \rightarrow V_2O_2 + 3K_2O$$

$$V_2O_3 + V \rightarrow 3VO$$

As it is a basic oxide, it dissolves in acids forming lavender coloured solution of hypovanadous salts having the ion $[V(H_2O)_6{}^{2+}$. These are unstable and powerful reducing agents. Solutions of these salts liberate hydrogen from water and deposit copper from solutions of cupric salts.

$$2V^{2+} + 2H_2O \rightarrow 2V^{3+} + 2OH^- + H_2$$

$$2V^{2+} + Cu^{2+} \rightarrow 2V^{3+} + Cu$$

If a solution of a hypovanadous salt is made alkaline precipitate of hypovanadous hydroxide is formed. This is a powerful reducing agent.

$$V^{2+} + 2OH^- \rightarrow V(OH)_2\downarrow$$

Vanadous Oxide, or Vanadium Trioxide, V_2O_3

It is obtained as black powder by heating vanadium pentoxide in hydrogen.

$$V_2O_5 + 2H_2 \rightarrow V_2O_3 + 2H_2O$$

It burns in air to form the pentoxide. It is basic, less than VO, and dissolves in acids with difficulty. Soultions of vanadous salts thus obtained are green. From these solutions a green precipitate of $V(OH)_2$ is obtained on making alkaline.

It reacts with thionyl chloride to form the trichloride.

Hypovanadous Oxide, VO_2 or V_2O_4

It is obtained by reducing the pentoxide the pentoxide with sulphur dioxide or fusing it with oxalic acid.

$$V_2O_5 + SO_2 \rightarrow SO_3 + V_2O_4$$

$$V_2O_5 + (COOH)_2 \rightarrow V_2O_4 + 2CO_2 + H_2O$$

It is an amphoteric oxide and with acids it forms vanadyl salts having VO^{2+} ion whereas with alkalis it forms hypovanadites which forms blue solutions.

$$V_2O_4 + 4HCl \rightarrow 2VOCl_2 + 2H_2O$$

$$2V_2O_4 + 2KOH \rightarrow K_2V_4O_9 + H_2O$$

Reaction of V_2O_4 and alkaline earth oxides at high temperatures in vacuum form vanadates of the types MVO_3, M_2VO_4 and M_3VO_5 (M = Ca, Sr, Ba).

Vanadium Pentoxide, V_2O_5

It is formed by heating ammonium metavanadate.

$$2NH_4VO_3 \rightarrow V_2O_5 + 2NH_3 + H_2O$$

It is also formed from the oxidation of lower vanadium oxides and hydrolysis of the oxohalide $VOCl_3$.

$$2VOCl_3 + 3H_2O \rightarrow V_2O_5 + 6HCl$$

It is a yellowish red powder and is sparingly soluble in water forming acidic soultions. With alkalis it gives a series of vanadates analogous to phosphate, eg, Na_3VO_{24} (ortho), $NaVO_3$(meta) and $Na_4V_2O_7$ (pyro). These compounds are crystallized at definite pH ranges – the tendency for condensation reaction increases with the decreasing value of pH.

$$VO_2(OH)_3^{2-} \rightarrow (VO_2)_2,(OH)_4^{2-} \rightarrow (VO_2)_3(OH)_6^{3-}$$

It undergoes reversible dissociation in the temperature range, 970 – 1250°C and this may partly be responsible for its catalytic properties.

$$2V_2O_5 \Leftrightarrow 2V_2O_4 + O_2$$

It dissolves in strong acids forming oxosalts having the ions VO^{3+} and $V(OH)^{4+}$.

Halides

A number of halides of vanadium are known.

Fluorides. Three florides are known.

(i) *Vanadium trifluoride, $VF_3 3H_2O$* : It is obtained at greenish yellow powder by careful concentrating the green solution obtained on dissolving vanadium trioxide in hydrofluoric acid. It is also obtained by prolonged heating of VCl_3 with dry HF. On heating in air, it forms vanadyl trifluoride, a white solid (m.p. 300°C).

$$2VF_3 + O_2 \xrightarrow{\text{heat}} \underset{\text{Vanadium oxytrifluoride}}{2OVF_3}$$

(ii) *Vanadium tetrafluoride, VF_4* : This is formed as a lime green powder by reacting anhydrous hydrogen fluoride with vanadium tetracahloride at low temperature. It dossolves in water to form blue solution in which it gets hydrolysed.

$$VCl_4 + 4HF \rightarrow VF_4 + 4HCl$$

It disproportionates at 327°C.

$$2VF_4 \rightarrow VF_5 + VF_3$$

It is a fluorinating agent, better than the pentafluoride.

(iii) *Vanadium Pentafluoride, VF_5* : It is the only pentahalide of vanaium and is formed by heating vanadium tetrafluoride in current of nitrogen at 600°C. The pentafluoride sublimes and is obtained as a white sublimate while a non-volatile residue of the trifluoride is left behind.

As it is extremely sensitive to moisture it first hydrolyses to vanadium oxytrifluoride and finally to the hydrated pentoxide.

$$VF_5 + H_2O \rightarrow VOF_3 + 2HF$$

$$2VOF_3 + 3H_2O \rightarrow V_2O_53H_2O + 6HF$$

It is a powerful fluorinating agent, better than, IF_5. It also forms addition compounds with suitable donor molecules, *e.g.*, $(XeF_6)_2VF_5$

Chlorides : Four chlorides of vanadium are known.

(i) Vanadium Chloride, VCl_2

It a obtained as a green residue by heating vanadium trichloride in a current of nitrogen at 800°C when the tetrachloride distils off.

$$2VCl_3 \xrightarrow{800^\circ C} VCl_2 + VCl_4 \uparrow$$

It dissolves in water forming a violet solution which soon evolves hydrogen and turns green (due to its oxidation by water). The violet solution possesses the ions $[V(H_2O)_6]^{2+}$.

(ii) Vanadium Trichloride, VCl_3

It is prepared by the action of HCl gas on the metal as well as by heating the solid. It is a violet solid.

It crystallizes from acidic solutions s the hexahydrate. The anhydrous salt is hygroscopic. It s solutions are acidic due to hydrolysis. It disproportonates on heating. It functions as Lewis acid.

(iii) Vanadium Tetrachloride, VCl_4

It is brown liquid. It is obtained by passing chlorine over heated vanadium or by passing vapours of $VOCl_3$ and Cl_2 over heated carbon.

$$V + 2Cl_2 \rightarrow VCl_4$$

$$2VOCl_3 + 2C + Cl_2 \rightarrow 2VCl_4 + 2CO$$

It is an unstable compound. It decomposes even at room temperature to form the trichloride and chlorine.

$$2VCl_4 \rightarrow 2VCl_3 + Cl_2$$

It is immediately hydrolysed by water.

$$VCl_4 + H_2O \rightarrow VOCl_2 + 2HCl$$

It is a Lewis acid and forms addition compounds with several donors.

Vanadium pentachloride is not known, but vanadium oxotrichloride, $VOCl_3$ may be obtained by passing dry chlorine over heated V_2O_3orV_2O_5 or heating a mixture of V_2O_5 and carbon in a current of chlorine.

$$2V_2O_3 + 6Cl_2 \rightarrow 4VOCl_3 + O_2$$

$$2V_2O_5 + 6Cl_2 \rightarrow 4VOCl_3 + 3O_2$$

$$V_2O_5 + 3C + 3Cl_2 \rightarrow \underset{\text{Vanadyl trichloride}}{2VOCl_3} + 3CO$$

$VOCl_3$, a lemon–yellow solid, is readily hydrolysed by water to form V_2O_5

$$2VOCl_3 + 3H_2O \rightarrow V_2O_5 + 6HCl$$

Vanadyl trichloride when heated with zine in a sealed tube or passed with hydrogen through a red hot tube forms $VOCl_2$ it is a bright green deliquescent substance. $VOCl_2$ is also formed by hydrolysis of VCl_4.

$$VCl_4 + H_2O \rightarrow VOCl_2 + 2HCl$$

VOCl. It is formed by heating $VOCl_2$ in vacuum. It is insoluble in water by dissolves in acids.

Bromides

Vanadium Dibromide, VBr_2 : It is formed as brown-orange crystalline plates by heating the tribromide with hydrogen. It resembles VCl_2 in properties.

Vanadium tribromide, VBr_3 : It is formed as a black solid by heating vanadium in a current of bromine. It is hygroscopic and the solutions are extensively hydrolysed. From acidic solution hexahydrate can be crystallized, Like VF_3 (but unlike VCl_3) it is thermally stable.

Vanadium tetrabromide, VBr_4 : It is formed by disproportionation of VBr_3. It is stable at –20°C It decomposes at room temperature to VBr_3 and Br_2.

The pentabromide is not known. Vanadyl tribromide VOBr3 may be obtained in an analogous manner to that VOCl in properties. It sublimes in vacuum at 327°C with extensive decomposition with bromine and the non-volatile VOBr. VOBr resembles VOCl in properties.

Iodides. These have been found to be less stable than the bromides which are less stable than the chlorides.

Vanadium diiodide, VI_2 : It is formed as rose-red haxagonal pates by heating the triodide, VI_3 under reduced pressure. It resembles other vanadium dihalides in properties.

It is used for preparation of pure vanadium.

Vanadium triiodide, VI_3 : It is a brown black solid. It is obtained by the direct combination of elements above 177°C. It decomposes on heating into the diiodide and iodine, otherwise it resembles VBr_3 in its properties.

Sulphides

The indentity of the pentasulphide, V_2S_5 is not fully established. It is formed when V_2S_3 and sulphur vapour mixture is heated at 400°C.

V_2S_3, vanadium sesquisulphide : It is the most stable sulphide of vanadium. It is formed by heating V_2O_3 in a stream of hydrogen sulphide, or by heating V_2O_3 in carbon disuphide vapour at 700°C. It does not dissolve in non-oxidizing acids but dissolves in nitric acid sometimes catching fire if the acid is concentrated.

When V_2S_3 is ignited in hydrogen, the mono-sulphide, VS is obtained.

$$V_2S_3 \xrightarrow{H_2, 1200^\circ C} 2VS + S$$

Vanadium sulphate, $V_2(SO_4)_3$: It can be prepared by the electroytic reduction of vanadium pentoxide in sulphuric acid. The green solution which results, deposits the green, lustrous needle of composition of $V_2(SO_4)_3$, $H_2SO_4.12H_2O$. On heating this compound at 180°C a yellow powder of $V_2(SO_4)_3$ is obtained and water SO_3 are liberated.

Vanadous sulphate is a powerful reducing agent. It forms alums with trivalent sulphates of iron, aluminium and chromium.

Uncommon Valence States of Vanadium : The valence state of –1 is illustrated by the rare compound $K_2[V(CN)_5NO]$. $Li[V(dipy)]4C_4H_8O$, and the compounds of the anion $[V(CO)_6]^{4-}$,

The zerovalent vanadium is present in vanadium hexa carbonyl $[V(CO)_6$, vanadium dipyridyl complex $[V(dipy)_2]$ and the complex $V[C_2H_4P(CH_3)_2]_3$. Vanadium hexacarbonyl has been prepared by treating vanadium tri-chloride VCl_3 in presence of diglyme, with carbon monoxide under a very high pressure at 20°C in presence of sodium a reducing agent. It is finally sublimed in presence of phosphoric acid.

$$VCl_3 + CO + Na \xrightarrow{\text{diglyme}} [Na(diglyme)_2][V(CO)_6]$$
$$\downarrow H_3PO_4, \text{sublime}$$
$$V[CO]_6$$

The valence state of +1 isillustrated by cation $[v(dipy)_3]^{1+}$ which forms salts with univalent anions like the halide ions and the nitrate ion.

Ammonium Vanadate, NH_4VO_3

Preparation

(i) It can be prepared by adding ammonia or a solution of ammonium salt to an aqueous solution of vanadium pentaoxide.

$$2NH_3 + V_2O_5 + H_2O \rightarrow 2NH_4VO_3$$

(ii) It can also be obtained by adding concentrated solution of ammonium chloride to a vanadate solution.

$$NaVO_3 + NH_4Cl \rightarrow NH_4VO_3 + NaCl$$

Properties

(i) Ammonium vanadate is a colourless crystalline solid.

(ii) It is moderately soluble (only about 6%) in water at room temperature.

(iii) On heating, ammonium vanadate decomposes into V_2O_5 and NH_3.

$$2NH_4VO_3 \rightarrow V_2O_5 + 2NH_3 + H_2O$$

(iv) Addition of dilute sulphuric acid to solutions of NH_4VO_3 gives a back red precipitate of V_2O_5.

$$2NH_4VO_3 + H_2SO_4 \rightarrow (NH_4)_2SO_4 + V_2O_5 \downarrow + H_2O$$

NIOBIUM OR COLUMBIUM

History

In 1801 Hatchett discovered a new oxide from columbite mineral. However, he could not isolate the element. In 1884 Rose found the existence of two elements in columbite, one was Ekeburg's tantalum and the other niobium itself. The name, niobium come from niobe, the mythological daughter of Tantalus. Since Hatchett had discovered the oxide some forty years earlier from columbite, the name columbium is still in use in America.

Occurrence

Niobium which is known as columbium always occurs associated with tantalum in nature. A number of ores have been discovered. Out of which the important ones are given below :

(i) *Columbite* : The general formula is (Fe, Mn) (Nb, Ta)$_2$O$_6$. It contains 26 – 77% Nb_2O_5 and 1 – 77% Ta_2O_5.

(ii) *Tantalite* : It also has the same general composition as the columbite. It may have 3 – 40% of Nb_2O_3 and 42 – 84% if Ta_2O_5.

(iii) *Pyrochlore* : It is the complex niobate of calcium and sodium with the general composition as (Na, Ca, Nb_2O_6F) It contains about 55% of Nb_2O_5 and 4% Ta_2O_5.

(iv) *Euxentie* : It is a complex niobate titanate of Y, Ca, Th, Zr, Er, and U. It possesses 18 – 35% on Nb_2O_5.

Fergusonite : It is metaniobate and tantalate with the composition (Y, Er, Ce) (nb, Ta) O_4.

Extraction : The main source for niobium or tantalum is the mineral niobite. Since the extraction for both the metals are similar, these two metals are extracted together, freed from other elements and then separated from each other. The methods followed for the extraction of the metal are only fusion methods since their compounds are insoluble in mineral acids except HF acid.

(i) *Fusion with caustic soda* : The ore niobite is finely powdered and then fused with NaOH.

$$Fe(NbO_3)_2 + 2NaOH \rightarrow \underset{\text{Sodium niobate}}{2NaNbO_3} + FeO + H_2O$$

$$Fe(TaO_3)_2 + 2NaOH \rightarrow \underset{\text{Sodium tantalate}}{2NaTaO_3} + FeO + H_2O$$

The fused mass is extracted with dilute HCl and the residue is digested with concentrated HCl. The acid solutions are mixed together which on concentration yields hydrated Nb_2O_5 and Ta_2O_5 as white powder.

$$Fe(NbO_3)_2 + 2HCl \rightarrow FeCl_2 + 5HNbO_3$$

$$2HNbO_3 \rightarrow Nb_2O_5 + H_2O$$

$$2NaNbO_3 + 2HCl \rightarrow 2NaCl + Nb_2O_5 + H_2O$$

$$2NaTO_3 + 2HCl \rightarrow 2NaCl + Ta_2O_5 + H_2O$$

The white powder is washed with water and then dissolved in hydrochloirc acid and then large amount of KF is added to the solution to produce potassium niobium oxyfluoride (K_2NMbOF_5) and potassium tantalum fluoride (K_2TaF_7).

$$Nb_2O_5 + 10HF \rightarrow 2NbF_5 + 5H_2O$$

$$NbF_5 + 2KF + H_2O \rightarrow K_2NbOF_5 + 2HF$$

$$Ta_2O_5 + 10HF \rightarrow 2TaF_5 + 5H_2O$$

$$TaF_5 + 2KF \rightarrow k_2TaF_7$$

Since the solubility of K_2NbO_5 is then times more than that of K_2TaF_5 most of the tantalum salt is separated out by fractional crystallisation from the niobium salt as needle shaped crystals.

(i) *Fusion with potassium hydrogen sulphate* : The ore niobite is finely powdered and fused with seven parts of potassium hydrogen sulphate in an iron vessel. The fused mass is extracted with boiling water when iron, manganese etc. form their salphates and pass into the solution. Tin and tungsten, if any are then separated by digesting the residue with yellow ammonium sulphide.

The residue containing the oxides of niobium, tantalum and silicon is washed with water and treated with hydrofluoric acid when most of the silicon dioxide is removed a silicon tetrafluoride vapours. Enough amount of potassium fluoride is then added to get K_2NbOF_5 and K_2TaF_7 as in the previous procedure.

$$2Fe(NbO_3)_3 + 4KHSO_4 \rightarrow 2FeSO_4 + 2Nb_2O_5 + 2K_2SO_4 + 2H_2O$$

$$Nb_2O_5 + 10HF \rightarrow 2NbF_5 + 5H_2O$$

$$Fe(TaO_3)_2 + 4KHSO_4 \rightarrow 2FeSO_4 + 2Ta_2O_5 + 2K_2SO_4 + 2H_2O$$

$$Ta_2O_5 + 10HF \rightarrow 2TaF_5 + 5H_2O$$

$$TaF_5 + 2KF \rightarrow K_2TaF_7$$

As in the previous method tantalum salt K_2TaF_7 is separated out from the solution by fractional crystallisation.

Chakravarty and Prince (1957) demonstrated that the above process is very effective in case of pyrochlore mineral also, which fails to respond to other conventional procedures.

(ii) *Chlorinaton* : Kroll (1956) used the chlorinaton procedure for opening the columbite and tantalite ores. In this process, the ore is heated with carbon in an are furnace. Most of the impurities like silica are removed leaving an infusible, metallic cake, which actually is mixture of two metals containing impurities. It is ground and digested with H_2SO_4, Fe Mn, Al etc are dissolved and are removed by filtration. The relatively pure carbide mixture is then chlorinated. Block (1958) found out that by maintaining the temperature to 1600° in above process and adding calculated amount of the carbon, the impurities of silicon, aluminium, magnesium and titanium can be converted into a very fluid stage, from which the viscous alloy of tantalum and niobium can be easily separated. This can be then subjected to chlorination at 500°.

Separation of tantalum and niobium : The two types of methods are as follows :

FRACTIONAL CRYSTALLISATION

This is oldest process known, based on the discovery of relative solubilities of K_2TaF_7 and K_2NbOF_5 H_2O. The method, though now largely replaced by solvent extraction methods, is still in use in some industries. The solution of the two above mentioned compounds, obtained at the end of above methods of the opening of the ores, is concentrated and allowed to crystallise. The less soluble K_2TaF_7 crystallises out and is separated by filtration.

The solution contains K_2NbOF_5 along with small amounts of tantalum, titanium and iron. Some K_2CO_3 is added to the boiling solutions which precipitates basic tantalum fluoride $Ta_2O_52K_2TaF_7$. This is filtered out and excess of sodium hydroxide is added; sodium niobate which is insoluble in presence of excess of NaOH is precipitated.

This is then brought again into solution phase; some potassium salt added and very pure potassium niobate obtained by fractional crystallisation. This when treated with hydrochloric acid, gives niobium oxide. Thus the end products are potassium fluotantalate (K_2TaF_7) and niobium oxide (Nb_2O_5).

Solvent Extraction Method

This method is more recent and extremely efficient. Hexone (MIBK) solvent has the unique property of dissolving only the fluorides of tantalum at low acidity and that of niobium at high acidity, leaving the impurities in aqueous phase, when a fluoride mixture solution, as one obtained at the end of first method of opening of ores is brought into its contact. The two then can be separately extracted again into aqueous phase. From these aqueous solutions, the respective hydroxides are precipitated by treating them with NH_4OH; the hydroxides, then, can be calcined to yield Nb_2O_5 and Ta_2O_5.

PREPARATION OF NIOBIUM METAL

1. *Sodium reduction of chloride :* Molten $NbCl_5$ is brought in contact with molten sodium. Reduction takes place and niobium metal granules sink to bottom.
2. *Carbide in reduction of pentoxide :* When a mixture of niobium carbide (NbC) and pentoxide (Nb_2O_5) is heated to temperature of 1800°C in vacuum, reduction of pentoxide into metal takes place and carbon monoxide is evolved. However some of the oxide is volatilized as such, causing loss in yield.
3. *Kroll's method :* $NbCl_5$ is sublimed over Mg in a glass tube at 750°C in the atmosphere of helium. After the completion of the reaction $MgCl_2$ is removed by distilling in vacuum at 900°C.
4. *Alumina thermic reduction of pentoxide :* Pure Nb_2O_5 and Al react when strongly heated to form nalloy from which aluminum can be volatilised during prolonged fusion in the electric vacuum furnace.

Purification of Metal

Van Arkel and de Boer's Method : The impure metal ad little niobium iodide is taken in an evacuated vessel through the centre of which runs a thin wire of very pure niobium which is heated to a very high temperature, while the temperature of the vessel is kept lower. Iodide decomposes on the wire, depositing the metal on it and releasing iodine which combines with the impure metal. This cycle continues till the wire becomes sufficiently thick and then it is taken out.

Use of Niobium

1. In the construction of nuclear reactor because of negligible tendency to absorb thermal neutrons and excellent high temperature mechanical properties.
2. As an alloying agent : It increases the strength of low carbon steel at high temperature and minimizes intergranular corrosion.

3. Pure niobium finds promising use in jet engines and rocket missiles.

Properties on Niobium

Physical

Niobium is a steel grey metal with a bright characteristic lustre and a light yellow tinge. The m.p. and b.p. are 1453° and 3700°C respectively. The specific gravity of the metal is 8.5.

Chemical

Niobium is a very reactive metal and resists the attacks of many chemical regents. The metal is difficult to obtain pure state because at high temperature it reacts with oxygen, carbon and nitrogen.

Action of Air

It is stable in air at ordinary temperature but on heating it forms a protective film of its oxide. Finely divide metal burns in air at about 400°C, forming Nb_2O_5 and when heated to red heat in oxygen it forms peroxide.

Action of Nitrogen

Nitrogen at 1200°C combines with niobium to form nitride.

Action of Water

It is unaffected by water but is decomposed a red heat liberating hydrogen.

COMPOUNDS OF NIOBIUM

Niobium is pentavalent in most of its compounds but it is trivalent in some of its compounds. These is also a tendency to form oxy-compounds.

Compounds of Pentavalent Niobium

Niobium Pentoxide

It is obtained by heating the element or sulphide or carbide or its nitride in air. It is more conveniently prepared by heating sodium or potassium niobate with an acid. The precipitate is hydrated oxide which on heating hives the pentoxide.

$$2KNbO_3 + 2HCl \rightarrow 2KCl + Nb_2O_5 + H_2O$$

$$Nb_2O_5.H_2O \rightarrow Nb_2O_5 + H_2O$$

Properties

(a) When obtained by precipitation method, it is a white amorphous powder, but if heated to red heat it becomes incandescent and passes into crystaline state.

(b) Nb_2O_5 is insoluble in water and is amphoteric but the acidic character is prominent. The compound metals at 1520°C and is reduced to lower oxides by H_2.

(c) Nb_2O_5 dissolves in fused KOH or $KHSO_4$

(d) The pentoxide can lose a large amount of oxygen by reduction with hydrogen without changing its crystalline structure till the composition of reduced product is above $NbO_{2.40}$

Other oxides like Nb_2ONbO and NbO_2 are known.

Niobium halides : Niobium pentahalides are prepared by heating niobium with halogens and they are prepared by heating the substances in the dry state since the are hydrolysed very rapidly by water.

Niobium Pentachloride : Niobium is obtained when niobium is heated in Cl_2 or when pentoxide Nb_2O_5 is heated with CCl_4.

$$2Nb + 5Cl_2 \rightarrow 2Nb_2Cl_5;\ Nb_2O_5 + 5CCl_4 \rightarrow 2NbCl_5 + 5COCl_2$$

$NbCl_5$ is a yellow crystalline solid and dissolves in many organic solvents. It is decomposed by water to form hydrated pentoxide.

It does not form double salt but forms addition compounds with organic compounds such as $NbCl_5.6C_5H_{11}N$ (Piperidine). $NbCl_5$ is completely miscible with $TaCl_5$ in the solid as well as in the metallic state.

Niobates and niobic acid : Niobates of the type M' $[NbO_3]$, M_4' $[Nb_2O_7]$ and M_7' Nb_5O_{16} are known. The nibates of alkali metals Li, Na and K and of bivalent metals of Fe, Mn and calcium viz. Li $[NbO_3]$, Fe $[NbO_3]_2$, Mn $[NbO_3]_2$ where first to be prepared.

COMPOUNDS OF TETRAVALENT NIOBIUM

Niobium Dioxide

NbO_2 has been obtained by heating niobium pentoxide to white heat in the atmosphere of H_2.

It is bluish black powder, insoluble in water, acids and alkalies. When heated in air it burns to form the pentoxide. It is a strong reducing agent at higher temperature.

Compounds of Trivalent Niobium

The sesqui-oxide Nb_2O_3 is not known and the formation of sesquisulphide Nb_2S_3 is slightly doubtful. The well known compound of Nb (III) is trichloride although the trifluoride has also been mentioned.

Compounds of Bivalent Niobium

The monoxide is only a simple compound which contains bivalent niobium. NbO is formed when niobium dioxide NbO_2 is heated to 1750°C in argon. It loses oxygen and forms a grey metallic powder.

Nobium Nitride

Is obtained by heating niobium with nitrogen at 1200°C. It is a grey metallic, interstitial compound. On strong heating in air it is converted into Nb_2O_5. Niobium also forms Nb_2N_2.

ZIRCONIUM

Discovery

Klaproth is 1789, while analysing samples of the mineral Zircon from Ceylon was able to isolate good quantities of a new oxide which closely resembled alumina (Al_2O_3) but differed from it in being insoluble in alkalies. Klaproth named the new element, zirconium from the word Zircon or Jargon Persian word which means gold coloured. Number of scientists such as Davy Vauquelin, Berzelius, Moissan tried to isolate the metal but got partial success, as they prepare impure metal. In 1857, Deville and Troost made vapour density determinations of the chloride and established the quadrivalency of the metal. Lely and Haburget in 1941 was also able to prepare pure zirconium.

Occurrence

Zirconium does not occur free in nature but is found widely distributed in nature in the form of oxide and silicate. It occurs in small amounts in earth a crust and rocks. The main sources of zirconium are :

Zircon or Hyacianth, $ZrSiO_4(55\%ZrO_2)$

Baddeleyite or Zirconia, Zr_2 $(80 - 95\%ZrO_2)$.

Large quantities of zircon comes from New south wales, India, Brazil and Australia. In India, zircon occurs in Bundelkhand and Hamirpur. Zirconium also occurs in the alluvial sands of Ceylon and Travancore. Some Volcanic leaves are having upto 50 per cent of zirconium.

Extraction

Form Zircon : The following methods are used for the extraction of zirconium from zircon.

1. Fusion with Alkali Hydroxides or Carbonates–Klaproth Method

Zircon mineral is fist of all powdered and then boiled with concentrated hydrochloric acid and washed with water. The residue is fused with about 5 times its weight of potassium hydroxide. The fused mass is then poured on a nickel sheet and cooled. The hot cake formed gets reached with water and filtered. The insoluble residue is having hydroxides of iron, zirconium and titanium and is dissolved in hydrochloric acid in which most of the residue dissolves, thus leaving a little silica. The clear solution is repeatedly evaporated with hydrochloric acid to dryness and the residue is extracted with water. Insolube fluosilicic acid *i.e.,* silica and hydrofluoric acid gets separated which is filtered off. The filtrate is made to treat with a little hydrochloric acid and zirconium oxychloride is crystallized out.

(i) $$ZrO_2 + 2KOH \rightarrow K_2ZrO_3 + H_2O$$

$$ZrSiO_4 + 4KOH \rightarrow K_2ZrO_3 + K_2SiO_3 + 2H_2O$$

(ii) $$2K_2ZrO_3 + 4H_2O \rightarrow 4KOH + 2ZrO_2 . H_2O$$

$$Zr(OH)_2 + 2HCl \rightarrow ZrOCl_2 + H_2O + H2.$$

(iii) When the finely pulverized zircon is fused with 4-6 times its weight of KHF_2, soluble potassium fluozirconate (K_2ZrF_6) and insoluble K_2SiF_6 are formed. The two are removed by filtration. The filtrate on cooling yields crystals of potassium fluozirconate.

$$ZrSiO_4 + 8KHF_2 \rightarrow K_2ZrF_6 + K_2SiF_6 \downarrow + 4KF + H_2O$$

(iv) In a recent method, the mineral zircon along with calcium oxide and carbon is heated in an electric furnace for a few minutes, during which most of the impurities get impurities get volatilized, where as the mineral yields the carbide.

$$ZrO_2 + 3C \rightarrow ZrC + 2CO$$

or $$ZrSiO_4 + 2CaC_2 \rightarrow ZrC + SiC + 2CaO + 2CO$$

After extracting the product with water and dilute HC, the residua is heated with chlorine at about 300°C to form zirconium tetrachloride which on treatment with ammonia gives precipitates of hydroxide. Zirconium hydroxide on heating yields zirconia.

$$ZrC + 2Cl_2 \rightarrow ZrCl_4 + C$$

$$ZrCl_4 + 4NH_4OH \rightarrow Zr(OH)_4 + 4NH_4Cl$$

$$Zr(OH)_4 \rightarrow ZrO_2 + 2H_2O$$

Removal of Iron

The separation of iron and titanium from zirconium has been widely studied. This can be carried out by :

(ii) The hydrolysis of zirconium sulphate whereby basic sulphate hydroxide is precipitated. It is used for large operations.

(iii) The action of water on an etheral solution of the respective chloride.

Flow Sheet

Zirocon
|
Powdered
|
Boiled with Cone. HCl and washed
|
Residue is fused with KOH
|
Fused mass poured on a nickel sheet and cooled
|
The hot cake leached with water and filtered
|

Filtrate — *Residue*

Hydroxides of Fe, Zr, Ti
| dissolved in
| HCl
Clear solution
Evaporated to dryness with HCl and the residue extracted with water

Residue : Silica etc : — Filtrate
Crystallized with a little of HCl
|
Zirconium oxychloride.

(iv) When sodium thiosulphate is added to a boiling solution of zirconium salt containing free acid, zircon, thoria and titanium are precipitated, leaving behind iron, aluminium, and rare earths.

(i) The action of sulphurous acid on neutral solution of their corresponding salts.

Purification of Zircon

Concentrated sulphuric acid is poured into a fairly concentrated aqueous solution of zirconium sulphate or chloride where upon a dense white crystalline precipitate of $Zr(SO_4)_2 . H_2O$ is formed. The precipitate is dissolved in equal weight of water and the hydrated sulphate is precipitated several times to obtain zirconium of high purity.

Best results are obtained by adding one volume of concentrated sulphuric acid to two volumes of concentrated zirconium solution and a small amount of HCl is to be added during recrystallization to prevent the precipitation of any iron that may be present as an impurity.

From Baddeleyite

(i) The powdered baddeleyite mineral is first of all boiled with HCl and then fused with alkali pyrosulphate or hydrosulphate in a graphite, cast-iron or fused silica crucible. The fused mass is now extracted with hot sulphuric acid until all the mass gets dissolved then cooled, poured into cold water and filtered.

From the filtrate zirconium can be precipitated out as phosphate which is then made to dissolve in hydrofluoric acid and the zirconium is reprecipitated as its hydroxide.

(ii) The ore is first of all finely pulverised and then fused either with caustic soda or soda ash (at high temperature) for converting into sodium zirconate (Na_2ZrO_3).

$$ZrO_2 + Na_2CO_3 \rightarrow Na_2ZrO_3 + CO_2$$

$$ZrO_2 + 2NaOH \rightarrow Na_2ZrO_3 + H_2O$$

The product (sodium zirconate) is sintered and extracted with water when zirconate gets hydrolysed to yield insoluble hydrated oxide, $ZrO_2 . H_2O$.

$$Na_2ZrO_3 + 2H_2O \rightarrow ZrO_2 . H_2O \downarrow + 2NaOH$$

Crystals of zirconyl chloride get dissolved in water and then treated with ammonium hydroxide to obtain hydrated zirconium oxide. The latter, on ignition, yields zirconia.

(iii) The ore is first of all finely powdered and then fused with 4 to 6 times its weight of KHF_2. The product is boiled with water, filtered when hot and the filtrate cooled to yield crystals of K_2ZrF_6.

$$ZrO_2 + 4KHF_2 \rightarrow K_2ZrF_6 + 2KF + 2H_2O$$

It is important to remember that owing to the lanthanide contraction, the size of hafnium and zirconium atoms and ions are almost identical, and so their properties too are nearly identical. Hence it becomes very difficult to separate the two elements which are always associated with each other in nature. Among the several methods used for the separation of lanthanides, ion exchange separation yields the best results. The chlorides dissolved in anhydrous methanol are allowed to pass through a column of silicagel. A 1.9 N solution of anhydrous HCl in methanol finds use for elution which preferentially carries forward the zirconium tetrchloride; and the hafnium tetrachloride is then made to elute with 7N sulphuric acid.

Metallurgy

During extraction, zirconium oxide, oxychloride or sulphate along with impuritie like iron, titanium, chromium, aluminium, and rare earths are obtained. It is possible to obtain the metal zirconium from these compounds by using the following methods :

1. Reduction of the Oxide

Zirconium may be obtained by reducing zirconium with carbon, calcium, magnesium or aluminium,. If carbon is used reduction is carried out at high temperature in an electric furnace.

$$ZrO_2 + 2C \rightarrow Zr + 2CO$$

Zirconium of 98 per cent purity can be obtained by the reduction of the oxide with calcium (Burget's Process)

$$2ZrO_2 + 2C \xrightarrow{600^\circ C} CaZrO_3 + Zr + CaO$$

$$CaZrO_3 + 2Ca \xrightarrow{1050^\circ C} Zr + 3CaO \cdot$$

2. Reduction of the Halides

It is also possible to prepare zirconium by reducing the zirconium tetrafluoride or tetrachloride vapour by passing over heated sodium, magnesium or aluminium.

$$ZrF_4 + 4Na \rightarrow 4NaF + Zr$$

$$ZrCl_4 + 2Mg \rightarrow 2MgCl_2 + Zr$$

$$3ZrCl_4 + 4Al \rightarrow 4AlCl_3 + 3Zr.$$

Kroll in 1955 obtained pure crystalline zirconium by the reduction of zirconium tetrachloride with molten magnesium in an atmosphere.

$$ZrCl_4 + 4Mg \rightarrow 2MgCl_2 + Zr$$

3. Reduction of Zirconfluorides

Berzelius Process : Zirconium is also obtained when potassium fluozirconate, K_2ZrF_6 is heated with potassium or sodium or aluminium.

$$K_2ZrF_6 + 4K \rightarrow 6KF + Zr$$

$$K_2ZrF_6 + 2Al \rightarrow 2AlF_6 + 2K + Zr$$

4. Van Arkel and De Boer's Filament Growth Process

It is possible, to prepare pure ductile zirconium by the thermal decomposition of zirconium tetraiodide, ZrI_4 by the filament growth process.

The method has been based on the fact that the decomposition temperature of iodide has been much lower than the volatilization temperature of zirconium. Hence it is deposited on the filament in crystalline form.

Apparatus is made of pyrex glass, having the shape. Zirconium iodide is made to introduce into the vessel through O. The opening O is now sealed and after evacuating the vessel S is also sealed. The tungsten filament F, stretched between the two tungsten rods T_1 and T_2 is then heated to about 1800°C by passing electric current, whereas the whole vessel gets heated to 600°C in an electric furnace. Zirconium iodide gets diffused to the incandescent filament and gets decomposed there, with the deposition of the metal.

$$\text{Impure Zr} + 2I_2 \xrightarrow[\text{on tungsten filament}]{1400^\circ C} Zr + 2I_2$$

This method finds use for producing metal of high purity used for research purposes. The metal can be further purified by zone refining when a heating coil is allowed to pass slowly along a tube containing the impure metal. The molten zone moves slowly and yields the pure metal equivalent to serval crystallizations. Pure solid separates and the impurities remain in the liquid and finish in the end zone.

5. Electrolytic Method

When zirconium tetrachloride mixed with ferric chloride is electrolysed using platinum wire cathode, zirconium get deposited on the platinum wire cathode.

Properties

Physical

(i) Zirconium (m.p 1875°, b.p. 3600°C), in the pure compact state, is a lustrous metal having a steel-like appearance. In the finely divided state, it forms a black powder.

(ii) It is known to exist in two forms, viz, crystalline and amorphous.

(iii) It is much softer than titanium.

(iv) It is malleable and ductile, when pure.

(v) Like titanium, it is corrosion resistant.

Chemical

(i) *Action of air :* When heated in air amorphous zirconium burns while the crystalline form resists the action of air at a red heat.

$$Zr + O_2 \rightarrow ZrO_2$$

The formation of oxide by air is able to explain its corrosion resistance property.

(ii) *Action of water :* Metallic zirconium does not get attacked by water. However, it is able to decompose steam to form zirconium hydroxide.

$$Zr + 2H_2O \text{ (steam)} \rightarrow Zr(OH)_2 + H_2$$

(iii) *Action of acids :* Although metallic zirconium is not affected by cold or hot hydrochloric acids, yet it gets attacked by hot 60% sulphuric acid an d a mixture of HF and HNO_3; aquaregia also attacks the metal vigorously. Hydrofluoric acid has been found to be the best solvent for this metal as well as its compounds.

$$Zr + 5H_2SO_4 \rightarrow H_2[Zr(SO_4)_3] + 2SO_2 + 4H_2O$$

$$3Zr + 4HNO_3 + 18HF \rightarrow 3H_2[ZrF_6] + 4NO + 8H_2O$$

$$3Zr + 4HNO_3 + 18HCl \rightarrow 3H_2[ZrCl_6] + 4NO + 8H_2O$$

$$Zr + 6HF \rightarrow H_2[ZrF_6] + 2H_2$$

(iv) *Action of non-metals :* When heated with hydrogen at about 700°C it forms a dihydride, ZrH_2. The gas also gets adsorbed by the metal especially when the latter is first of all heated at 1000°C with hydrogen and then cooed. It reacts with nitrogen above 1000°C and yield s the refractor nitride, ZrN. The metal is also attacked by gaseous HCl and CL_2 at a dark red heat. With halogens it yield the corresponding tetrahalides.

$$Zr + 2Cl_2 \rightarrow ZrCl_4$$

$$Zr + 2Br_2 \rightarrow ZrBr_4$$

$$Zr + 2I_2 \rightarrow ZrI_4$$

(v) *Action of alkalies :* Very pure zirconium is not acted upon by alkali but the ordinary zirconium evolves hydrogen. The amorphous zirconium is slowly oxidised and dissoved by fused alkalkies.

$$Zr + 2NaOH + O \rightarrow Na_2ZrO_3 + H_2$$

Uses

1. *Zirconium alloys :*Alloys of zirconium are of great interest. Cu-Zr alloys have almost constant specific resistance while the tensile strength and low resistance are its distinct advantages. It imparts to steel toughness, tenacity. Resistance to wear and resistance to corrosion. It is also employed as a scavanger in the steel industries.
2. *Refractory material :* Zirconium oxide is used in the manufacture of highly refractory wares and crucibles. Fire bricks are also made with zirconia and clay.
3. *Atomic reactors :* Metallic zirconium is now becoming an element of outstanding interest in the field of atomic energy because of its very slow tendency to absorb slow neutrons. At present its use in nuclear reactions is being investigated.
4. *Glass wares :* When zirconium oxide is added to glass, its resistance to chemical reagents increases but the viscosity, density and thermal expansion decreases.

5. *Electrical appliances :* It finds extensive applications in photoflash bulbs, radio-transmitter tubes, vacuum tubes, electric welding etc.
6. *Catalyst :* It is also suggested as a catalyst in hydrogenation an din the synthesis of ammonia.
7. In the form of powder it is used for ammunition, detonators and the preparation of smokeless flash-light powders.

Compounds of Zirconium

Zirconium, being amphoteric forms two series of compounds zirconium salts and zirconates. It has also a great tendecy to form zirconyl salts. Normal salts are hydrolysed by water.

Zirconium forms essentially quadrivalent compounds. It is bivalent in the hydride ZrH_2.

Oxides : The oxide of zirconium which are known are ZrO, Zr_2O_2, ZrO_2, Zr_2O_3 and ZrO_3. Of all these ZrO_2 has been widely studied.

TANTALUM

History

In 1802 when Epeberg was studying two new minerals, one form Finlnad and other from Sweden, he found the same new element in both these minerals. He gave the name tantalum to new element because of tantalizing difficulties he encountered while dissolving the minerals in acids. In 1844 Rose found that both niobium and tantalum are present in columbite.

Occurrence

Tantalum occurs associated with niobium ores. A great many tantalum minerals have been described. There are :

1. Tantalite. $Fe(TaO_3)_2$, 86% TaO.
2. Pamarspite . Fe (Fe, Ca, $UO_2)_3$ (Ce, $Y)_2$ (Nb, $Ta)_6O_2$
3. Fergusonite. (Y, Er, Ce (Nb) ; Ta) O_4 : 4–43%

Extraction

Extraction of the tantalum metal has been already given under niobium as both tantalum and niobium occur in the same minerals and ores are also extracted by the same procedures

Extraction of Tantalum

1. *Potassium reduction of faluotantalate* : The following reduction is carried out in a steel bomb at very high temperature.

$$K_2TaF_7 + 5K \rightarrow 7KF + Ta.$$

The fluoride of potassium is leached but with water, leaving a very fine pure metal powder.

2. *Carbide reduction of Ta_2O_5* : The process is the same as described in case of niobium. The temperature required is 2000°C.

Electrolysis of potassium fluotantalate : Pure tantalum is obtained by electrolysis of a molten mixture of K_2TaF_7 and Ta_2O_5 at 900°C in a cast iron vessel which acts as cathode. Anode consists of graphite rods. The metal formed settles down at the bottom of the vessel. The metal formed by anyone of the above processes is further purified by inert atmosphere are melting or electron beam melting .

Uses of Tantalum

1. In the construction of nuclear reactors because of negligible tendency to absorb thermal neutrons and excellent high temperature mechanical properties.
2. In the production of corrosion resistant chemical equipments.
3. In the manufacture of high temperature vacuum tubes.
4. As a substitute for platinum in pen-nibs.
5. In the manufacture of surgical and dental requirements because of its excellent mechanical properties of resistance to chemicals.
6. It is also used in rectifiers for A.C. condenser and radar equipment.

Properties of Tantalum

Physical

Tantalum is a white grey metal with high specific gravity. It melts at 3000°C. Pure metal is malleable, ductile and tough. Pure tantalum is a soft metal but becomes harder when mixed with impurities.

Chemical

(a) *Action of air* : Tantalum is stable in air at ordinary temperature. Powdered metal when heated in air to red heat, burns forming Ta_2O_5.

(b) *Action of water :* Red hot metal decomposes steam rapidly and hydrogen evolved catches fire.

(c) *Action of halogens :* Tantalum metal heated 250°C burns in chlorine to form the pentachloride. Vapours of Br_2 attack tantalum at high temperature to form the pentabromide but I_2 does not act upon it.

(d) *Action of hydrogen, carbon and nitrogen :* When Ta is heated in nitrogen, it forms nitride TaN. Ta forms two carbides TAC and Ta_2C which are hard solids while with hydrogen Ta_2H and TaH are obtained.

(e) *Action of alkalies :* Tantalum dissolve in fused alkalie to form metal tantalates.

(f) *Action of acids :* with the exception of hydrofluoric acid Ta does not dissolve in boiling acids and in aqua regia. It dissolves in HF to form the acid $HTaF_6$.

$$2Ta + 12HF \rightarrow 2HTaF_6 + 5H_2$$

(g) *Readcing action :* Tantalum powder reduces many oxides such as $P_2O_5.SO_2PbO_2$ and mercuric chloride to the element state.

$$4Ta + SO_2 \xrightarrow{500^0} 2Ta_2O_5 + 5S$$

$$P_2O_5 + 2Ta \xrightarrow{\text{heated to}} 2P + 5Ta_2O_5$$

$$5Hgcl_2 + 2Ta \xrightarrow{\text{High temp.}} 2TaCl_5 + 5Hg$$

Action of metals Na and K do not alloywith Ta, while Mo and Cr an easily soluble in molten Ta.

Compounds of tantalum : Tantalum compounds are in general similar to those of niobium.

Penta-valent Tantalum

Tantalum Pentoxide : It is obtained by treating sodium or potassium tantalate with an acid. The precipitate is an hydrated oxide which on heating forms the pentoxide.

$$2KTaO_3 + 2HCl \rightarrow 2KCl + Ta_2O_5 + H_2O$$

$$Ta_2O_5.H_2O \rightarrow Ta_2O_5 + H_2O$$

Tantalum pentoxide is a white amorphous powder insoluble in water and in acids except HF.Ta_2O_5 is insoluble in alkali solution but on fusion with alkalies it forms alkali tantalates. It is not reduced by H_2 and is not attacked by Cl_2, Br_2, HCl, etc. even at high temperatures.

Tantalum also forms an oxide, TaO_2

Tantalum halides : Ta forms pentahalides with all halogen sand they obtained by heating the substances in dry state.

It forms number of double fluorides.

Tantalum pentachloride : It is obtained by heating tantalum pentoxide (Ta_2O_5) and carbon in a stream of chlorine. It may also be formed by heating the element or its carbide, nitride in Cl_2. It is a white crystalline solid which is easily volatilized. It is hydrolysed by water giving tantalic acid and HCl.

$$2TaCl_5 + 6H_2O \rightarrow Ta_2O_5.H_2O + 10HCl$$

Tantalates and tantalic Acid : When Ta_2O_5 is fused with excess of Na_2CO_3, the quantity of CO_2 evolved shows the formation of orthotantalate (Na_3TaO_4) but after washing the residue with cold water insoluble residue is made of metal tantalate ($NaTaO_3$). The properties of meta tantalate are similar to those of meta-niobates. The alkali metal tantalates are easily hydrolysed in solution by the addition of dil H_2SO_4. The precipitate of hydrated pentoxide is known as 'tantalic acid'.

Compounds of Tetravalent Tantalum

Tantalum dioxide : This oxide is obtained by the ignition of the Ta_2O_5 with carbon.

The dioxide is grey solid which is not attacked by acids but dissolves in NaOH and wen it is heated in air, it is oxidised to pentaoxide Ta_2O_5. All four halides of Ta(IV) are known and are obtained by the reduction of MX_5.

Compounds of Trivalent Tantalum

Tantalum trichloride : $TaCl_3$ is formed as a green solid by the reduction of $TaCl_3$ with H_2 or aluminium powder. When heated at 500°, it changes into $TaCl_2$ and $TaCl_5$.

$$3TaCl_3 \rightarrow 2TaCl_2 + TaCl_5$$

Tantalum trichloride dissolves in water forming green solution. When alkali solution is added to the aqueous solution of $TaCl_3$ a green hydroxide $Ta(OH)_3$ is precipitated. No oxide of trivalent Ta is known.

Compounds of Bivalent Tantalum

Tantalum dichloride : $TaCl_2$ is the only well characterised red compound of bivalent tantalum although the sulphide TaS has been mentioned.

Tantalum dichloride is obtained as the product formed by decomposition of trichloride, $TaCl_3$ at 500–600°C. It is a green solid at ordinary temperature. It is very slightly soluble in water and decomposed even by cold water with the formation of trivalent tantalum and hydrogen.

$$TaCl_2 + H_2O \rightarrow Ta^+ + 2Cl^- + OH^- + 1/2H_2$$

Note : TaO is not known.

GERMANIUM

Discovery

Germanium was predicted by Mendeleeff (1871) as Kea-silicon. Clemens Winker, a German chemist discovered the actual element in 1886 from its ore argyrodite and studied its properties. He gave it the name Germanium in honour of his country Germany.

Occurance

Germanium is a scarce element it occurs in nature in very few minerals like

(1) Argyrodite, $kAg_8[GeS_6]$ or $4Ag_2S.GeS_2$.

(2) Germanite, $7Cu.FeS.GeS_2$.

(3) Confieldite, $4Ag_2S(Ge - Sn)S_2$.

Other sources of germanium are coke furnace flue dust and gas liquor. Residues from zinc and lead smelters are sometimes found rich in germanium

Extraction

(1) From Argyrodite : Winkers Method

It is first of all powdered and then fused with sodium carbonate and sulphur. The fused mass is extracted with water and the solution treated with dilute sulphuric acid so as to decompose completely the sodium sulphide present. It is kept for a day and then filtered to get rid of sulphides of arsenic and antimony and the sulphur. The filter is treated with HCl acid and saturated with hydrogen sulphide. A white voluminous precipitate of germanium sulphide obtained is washed and roasted with HNO_3 and then ignited to yield GeO_2.

(2) From Germanite

The coarsely ground dry germanite is heated to 350°C in a an atmosphere of chlorine in a chlorination chamber when volatile chlorides of Ge, Si, As and S are distilled out. Now if the chlorides are subjected to fractional distillation where by germanium tetrachloride, $GeCl_4$ is distilled over at 83°C and could be separated from $AsCl_3$ (B.P. 130°C) and S_2Cl_2 (B.P. 138°C). The clear red distillate is poured over crushed ice when germanium chloride hydrolyses yielding a light yellow hydrated germanium dioxide as a suspension.

Now, the hydrated germanium dioxide is saturated with Cl_2, distilled and hydrated with ice. The white ppt. obtained is dried and ignited to yield GeO_2. Germanium can be recovered form wash solutions. This process is very efficient and germanium obtained is spectroscopically pure.

Metallurgy

(1) Winkler's Method

Germanium dioxide can be reduced to the metal by heating to red heat in a current of hydrogen.

$$GeO_2 + 2H_2 \xrightarrow{540^\circ C} Ge + 2H_2O$$

(2) Tressler and Dennis Method

Germanium metal can be obtained by reducing germanium dioxide with carbon under a reflex of sodium chloride.

$$GeO_2 + 2C \rightarrow Ge + 2CO$$

(3) Reduction with Aluminium

Germanium dioxide may also yield white lustrous metal by heating it with metallic aluminium.

$$3GeO_2 + 4Al \rightarrow 3Ge + 2Al_2O_3$$

Properties

Germanium occurs as a greyish-white, brittle metal having a conchoidal fracture and crystalline structure. The crystals are octahedral.

Its Sp. Gr. is 5.469 as 20°C and m.p. is about 958°C. It is diamagnetic. The electrical conductivity of germanium lowers as the temperature gets decreased. (It is a semi-conductor).

Action of Heat

At an incipient red heat, the metal is coated with a thin film of the dioxide.

Action of H_2

It does not react with hydrogen but when heated above 700°C it yield a volatile hydride

Action of Water

Water has no action on the metal. Steam is decomposed by red hot element.

$$Ge + 2H_2O \rightarrow GeO_2 + 2H_2$$

Action of Acids

The metal slightly dissolves in hot hydrofluoric acid. It is insoluble in hydrochloric acid and dilute sulphuric acid. The action of nitric acid is very violent and hydrated germanium dioxide and nitric oxide are formed.

$$2Ge + 2HNO_3 \rightarrow 2GeO_2 + 2NO + H_2$$

If heated with conc. H_2SO_4, it evolves sulphur dioxide.

Action of Alkalies

Germanium does not get attacked by dilute alkalies, but it readily dissolves infused alkali hydroxide, sodium carbonate, sodium peroxide and potassium nitrated to yield germanate ($GeO^{2-}{}_3$). It is also readily attacked buy an alkaline solution of hydrogen peroxide.

Action of Halogen

Germanium burns with a bluish white flame when heated in chlorine yielding germanium tetrachloride.

The metal also burns in bromine with yellowish flame yielding the tetrabromide and with iodine at red heat to form iodide.

Uses

(1) It finds and extensive use in the field of electronics. It finds use incurrent rectifiers of radar and is of increasing importance in commercial microwave application.

(2) It also finds use on television receivers, transistors and appliances for the transmission of infra-red rays.

(3) It finds use in germanium resistance thermometers.

(4) It finds use in making alloys. Small quantity of germanium to aluminium alloys as duralumin increases the strength of the alloy and improves its engineering properties. Copper can be hardened by the addition of small quantities of germanium.

(5) Germanium dioxide finds use in marking glass.

GENERAL DISCUSSION OF TITANIUM

The elements of titanium group-titanium, zirconium and hafnium-constitute group IVB of the periodic table. These are d-block elements having the configuration (n – 1) d^2, ns^2 as given in the following :

Table 1

Element	*Symbol*	*At. No.*	*Electronic configuration*	*Oxidation States*
Titanium	Ti	22	[Ar] $3d^2$ $4s^2$	(–2), (–1), (0), (2), 3, 4
Zirconium	Zr	22	[Kr] $4d^2$ $5s^2$	0, (2), (3),4
Hafnium	Hf	72	[Xe] $4f^{14}$ $5d^2$ $6s^2$	3, 4

General Characteristics

These are all metals with high melting and boiling points. Some physical constants of this group are given in Table 1.

(1) Oxidation States

Most common and most stable oxidation state of these elements is +4. Since energy required to remove four electron from an atom is large, their anhydrous compounds like $TiCl_4$ are covalent. These are formed by sharing of two s and two d electrons with other atoms. The covalent compounds so obtained are tetrahedral. In +4 oxidation state, the elements have a d° configuration and these are, therefore, diamagnetic and colourless or white. The oxides MO_2 are regarded ionic but these are insoluble and there are no M^{4+} ions in solution.

Table 2 : Some physical properties of titatanium group

Property	*Titanium*	*Zirconium*	*Hafnium*
Atomic Mass	47.88	91.22	178.49
Metal radius/pm	147	160	159
Ionic radius/pm 6-coordinate, M(IV)	60.5	72	71
Melting point/K	1940	2130	2495
Boiling point/K	3558	4473	4723
Density (25°C K)/gcm^{-1}	4.50	6.51	13.28
Electronegativity	1.5	1.4	1.3
DH_{fus}/KJ mol^{-1}	18.8	19.2	1257
DH_{vap}/KJ mol^{-1}	425	567	571
Electrical resistivity (25°C)/μ ohm cm	42.0	40.0	35.1

Compounds in +3 oxidation state are reasonably stable and reducing agents. For example Ti^{3+} ions ar more strongly reducing than Sn^{2+} ions. M^{3+} ions have a d^1 configuration and are paramagnetic due to the presence of one unpaired electron.

There being only one d electron, only d-d electronic transition is possible. This can give only band in visible spectrum and we find nearly all compounds of these elements are purple coloured.

The +2 oxidation state is very unstable and so strongly reducing that it reduces water. Thus, there are only a few compounds in solid state in +2 oxidation state.

$$2Ti^{2+} + 2H_2O \rightarrow 2Ti^{3+} + H_2 + 2OH^-$$

Lower oxidation states disproportionate into higher and lower oxidation states.

$$2Ti(III)Cl_3 \xrightarrow{\text{Heat}} Ti(IV)Cl_4 + Ti(II)Cl_2$$

$$2Ti(II)Cl_2 \rightarrow Ti(IV)Cl_4 + Ti(0)$$

The zero, –1 and –2 oxidation states are formed in compounds like $[Ti^{o}(dipy)_3$, $Li[Ti^{2-}(dipy)_3]$ 3.5 THF and $Li_2[Ti^{2-}(dipy)]$ 5THF (where THF = etraudrpfiran).

(2) Size of Atoms and Ions

The covalent and ionic radii increase from Ti to Zr but there in no increase from Zr to Hf. This is on account of lanthanide contraction occurring because of the filling of 4f level.

(3) Reactivity and Passive Behaviour

The metals are passive or unreactive at low temperature due to the formation of a protective oxide film on the surface of the metal. Acids and alkalis have no reaction at room temperature. Titanium dissolves in hot conc. HCl giving Ti^{3+} ions and in concentrated nitric acid forming TiO_2 nH_2O_2. Zirconium dissolves in HF. This is due to their tendency to form hexafluoro complexes.

$$Ti + 6HF \rightarrow H_2[TiF_6] + 2H_2$$

At high temperatures (above 600°C) the metals are very reactive to form compounds of the type MO_2, MX_4, MN, MC (interstitial nitrides and carbides) by direct combination.

Powdered metals adsorb hydrogen to give interstitial hydrides of formula MH_2. Amount of hydrogen adsorbed depends on temperature and pressure. Hydrides are stable in air and do not react with water.

(4) Nature of Oxides

The oxides of the formula MO_2 are stable, non-volatile and insoluble. On ignition. they becomes refractory and ar used as white pigment. Titanium oxide is amphoteric but more acidic than basic, but ZrO_2 and HfO_2 are basic. TiO_2 dissolves in both acids and alkalis forming titanyl compounds and titanates respectively.

$$\underset{}{TiO_2} + \underset{(conc.)}{H_2SO_4} \rightarrow \underset{\text{Titanyl sulphate}}{TiOSO_4}$$

$$TiO_2 + \underset{(conc.)}{2NaOH} \rightarrow \underset{\text{Sodium titanate}}{Na_2TiO_3.nH_2O}$$

The reaction of TiO^2 with acids does not give Ti^{4+} ions in solution because of the high charge/radius ratio.

The titanyl ion polymerizes to form $(TiO^{2+})_x$ chains.

Similarly, Zr^{4+} ions do not exist in solution but polymerized ZrO^{2+} (zirconyl) ions are obtained . Anhydrous titanates Na_2TiO_3 and $Na_2Ti_2O_5$ are obtained by fusing TiO_2 with NaOH or Na_2CO_3. Calcium titanate $CaTiO_3$ occurs naturally and has a perovskite structure. The $CaZrO_3$ is isomorphous with perovskite and is a mixed $CaO.ZrO_2$

When a neutral titanium (IV) solution is treated with H_2O_2 and alcohol, a yellow compound corresponding to TiO_3. $2H_2O$ is obtained. The compound is an uncharged species and liberates iodine from an iodide solution corresponding to the presence of one peroxy group in the molecule. Zirconium and hafnium also form such compounds under similar conditions. The compound is formulated as $M(OH)_3$. OH (where M = Ti, Zr or Hf). Treatment of the above titanium or zirconium compound with alkaline H_2O_2 gives tetraperoxytitanate (IV) or tetraperoxyzironate (IV) ion

Tetraperoxytitanate (IV) zirconate (IV) ion (where M = Ti or Zr)

When an acidified titanium (IV) solution is treated with H_2O_2 an intense yellow-orange colour is obtained. The colour reaction is used for detection and colorimetric determination of both titanium (IV) and H_2O_2. The species produced contains one bidenatae peroxy group per titanium atom.

$$TiO(SO)_4 + H_2O_2 \rightleftharpoons \underset{\text{yellow-orang colour}}{Ti(O_2)SO_4} + H_2O$$

The compound $Ti(O_2)SO_3 3H_2O$ and similar compounds containing the anions of other acids have been isolated.

(5) Halides

Tetrahalides, MX_4 are formed by interaction of the metal with halogen. The best known titanium tetrachloride, $TiCl_4$ is formed by heating TiO_2 with carbon and chlorine. The other halides can be prepared from $TiCl_4$ by reaction with appropriate hydrogen halide.

$$TiCl_4 + 4HX \rightarrow TiX_4 + 4HCl$$

The halides are vigorously hydrolysed by water and fume in moist air.

$$TiX_4 \xrightarrow{H_2O} TiOX_2 \xrightarrow{H_2O} TiO_2.nH_2O$$

The compounds act as acceptor to a wide variety of donor ligands such as F^- ions, Cl^- ions, phosphines R_3P, arsines R_3As, ammonia, pyridine and trimethylamine

$$TiF_4 + 2F^- \rightarrow [TiF_6]^{2-}$$

$$TiCl_4 + 2Cl^- \rightarrow [TiCl_6]^{2-}$$

In general the complexes obtained have octahedral structure. However, a few unusual five coordinate (such as $TiCl_4.AsH_3$), seven coordinate (shown in Fig. 3.2, a and b) and eight-coordinate (shown in Fig. 3.2 c and d) complexes are also known.

(7) Compounds with Hydrogen Carbon and Nitrogen

The hydrides of titanium, zirconium and hafnium are brittle metal-like black or grey powders having composition corresponding to MH2. These compounds are intermediate between ionic hydrides and solid solutions. Carbides and nitrides are formed by direct interaction at high temperatures. MC and MN compound (of various compostions) are hard, crystalline solids, chemically inert and have refractory properties.

(8) Aqueous Chemistry

Oxidation state (IV). The hydrated ions, M^{4+} cannot exist in solution because of the high charge-to-radius ratio and are strongly hydrolysed. Normal salts such as $Zr(NO_3)_4.5H_2O$ can only be isolated if the solution is sufficiently acidic. Numerous oxo-metal (IV) salts have been isolated but the existence of MO^{2+} ions has not been proved.

Thus, $TiOSO_4H_2O$ Contains chains of -Ti-O-Ti-O- with each Ti octahedrally coordinated to two bridging oxygen atoms, one water molecule and one oxygen atom from each of three sulphate ions. $ZrO(NO_3)_2$ is also an oxygen bridged chain. There is more extensive aqueous chemistry of zirconium; because of its larger size it has a lower tendency towards complete hydrolysis.

There is no positive evidence of the existence of Zr^{4+} or ZrO^{2+} ions in solution. In 1–2 M perchloric acid the only major species present are $[Zr_3(OH)_4]^{8+}$ and $[Zr_4(OH)_8]^{8+}$.

Oxidation state (III) is clearly expressed only in titanium. Aqueous solutions of Ti (III) are violet in colour and are reducing in charter. This is used frequently as a reducing agent. In acid medium the main species present is $[Ti(H_2O)_6]^{3+}$ which hydrolyses as follows :

$$[Ti(H_2O)_6]^{3+} \rightarrow [Ti(OH)(H_2O)_5]^{2+} + H^+$$

Only Ti (II) is known. It has no aqueous chemistry because it is oxidized in water.

(9) Organometallic Compounds

Organo-titanium compounds assumed importance mainly because of the discovery by Ziegler and Natta (around 1955) that ethylene can be polymerized in presence of a mixture of $TiCl_4$ and $AlEt_3$ in a hydrocarbon solvent at room temperature and atmospheric pressure.

Group IVB elements form few alkyl and aryl compounds, which are unstable to both air and water. MMe_4 can be prepared by the reactions of LiMe and MCl_4 in ether at low temperature. The yellow titanium compound, $TiMe_4$ and the red zirconium $ZrMe_4$ decompose to the metals at temperatures above 253°K and 258°K respectively.

Cyclopentadienyls are, however, much more stable and common, and are known for M(IV), M(III) and M(II). The compounds $M(C_5H_5)_4$ are prepared from MCl_4 and NaC_5H_5. Another type of compounds with the composition $[M(C_5H_5)_2X_2]$ are known and are prepared by the action of NaC_5H_5 on MCl_4. $[M(C_5H_5)_2]$ are the only well characterized carbonyls formed by group IVB elements.

5

Basic Properties of Iodine

INTRODUCTION

Brickenbach and Kellermann : defined the term 'pseudohalogens' as:

"A large number of inorganic radical are known which in the free state possess properties similar to those of the elemental halogens and in the form of anions possess similar to those of the halide ions."

Browne and his co-worker suggested the name 'halogenoids' instead of pseudo-halogens. Later, Walden and Audrieth defined halogenoid :

"Any univalent chemical aggregate composed of two or more electronegative atoms which shows in the free state certain characteristics of halogens and which combines with hydrogen to form an acid and with silver to form salt insoluble in water."

Similarly pseudohalides may be defined as follows :

There are certain uninegative ions (*e.g.*, CN^-, OCN^-, N^-_3 etc.) Which are made up of two or more electronegative atoms and resemble halide ions in some respects. These uninegative ions are called pseudohalide ions.

Examples of some important pseudohalide ions are : cyanide ion (CN^-), isocyanide in (NC^-), fulminate ion ONC^-), cyanate or oxycyanite ion (OCN^-), thiocyanate ion (SCN^-), isothiocyanate ion (NCS^-), tellurocyanate ion ($TeCN^-$), selenocyanate ion ($SeCN^-$), isoselenocynate (ionic compounds) given by the pseudohalide ions [*e.g.*, AgCN, $Pb(CNS)_2$ etc.] are called pseudohalides.

As the dimers of halide ions are called halogens, the covalent dimers of the pseudohalide ions are called *pseudohalogens* or *halogeniods*. So far only a limited number of pseudohalogens have been isolated and characterised. Examples of some pseudohalogens or halogenoids are : cyanogen, $(CN)_2$; azidocarbon disulphide, $(SCSN_3)_2$.

The above examples of halogenoids and halogenoid ions are grouped in a tabular form :

Table 1

Free Halogenoid	*Halogenoid ions*
Cyanogen $(CN)_2$	Cyanide CN^-
Thiocyanogen $(SCN)_2$	Thiocyanate SCN^-
Selenocyanogen $(SeCN)_2$	Selenocyanate $SeCN^-$
Telluro cyanogen $(CNTe)_2$	Tellurocyanate $TeCN^-$
Azido carbon disulphide $(SCSN_3)_2$	Azido dithiocarbonate $SCSN^-_3$
Oxycyanogen $(OCN)_2$	Cyanate OCN^-
	Fulminate OCN^-
	Azide N^-_3

Oxycyanogen is not available in the free state.

THE PSUEDOHALOGENS AND THE HALOGENS

Similarities between the halogenoids and halogens are striking . Some similarities are discussed below :

(i) Volatile and dimeric nature. Like halogens, pseudohalogens are dimeric and are fairly volatile in the free state with the exception of polymeric thiocyanogen.

(ii) Isomorphous nature. They are often isomorphous when free and in the state. For example, Cl_2 is isomorphous with $(CN)_2$ and similarly Br_2 is isomorphous with $(SCN)_2$.

(iii) They combine with many metals to give salts. For example, silver and mercury salts of psudohalogens are like the corresponding halides (AgCl, AgBr, AgI)

(iv) With hydrogen, they form monobasic acids which are, however very weak compared with halogen acids. For example,

$$pK_a \text{ for } N_3H = 4.4$$

pK_a for HCN = 8.9

(v) They form compounds among themselves, such as CNN_3 which are analogous to interhalogen compounds (ICl).

(vi) They form double and complex salts like $K_2[Hg(SCN)_4]$ $NaAg(CN)_2$ which are analogous to the corresponding halides.

(vii) The pseudo-halogens may be prepared in the free state either by chemical or electrolytic oxidation of the hydracids or their salts or by thermal decomposition of higher valent compounds.

(viii) Compounds of halogens and pseudo-halogens are very similar. For example :

Pseudohalogen	ICN	$Si(NCS)_4$	$pb(SCN)_4$	Co(N3)2
Halogens	ICl	SiC14	pbC14	CoC12

(ix) Pseudohalogens react with alkalies in rather a similar manner to the halogens :

$$(CN)_2 + 2OH^- \rightarrow CN^- + OCN^- + H_2O$$

$$(SCSN_3)_2 + 2OH^- \rightarrow SCSN^-_3 + OSCSN^-_3 + H_2O$$

$$Cl_2 + 2OH^- \rightarrow Cl^- + OCl^- + H_2O$$

(x) Like halogens, pseudohalogens form addition compounds with unsaturated hydrocarbons.

$$\begin{matrix} CH_2 & & SCN & & CH_2SCN \\ \| & + & | & \rightarrow & | \\ CH_2 & & SCN & & CH_2SCN \end{matrix}$$

$$\begin{matrix} CH_2 & & Cl & & CH_2Cl \\ \| & + & | & \rightarrow & | \\ CH_2 & & Cl & & CH_2Cl \end{matrix}$$

(xi) Certain lead (IV) compounds of pseudohalogens are decomposed in a similar manner to the halogens.

$$Pb(SCN)_4 \rightarrow Pb(SCN)_2 + (SCN)_2$$

$$PbCl_4 \rightarrow PbCl_2 + Cl_2$$

(xii) The hydracids of halogens and pseudohalogens react with oxidising agents in a similar manner.

$$MnO_2 + 4H^+ + 2SCN^- \rightarrow Mn^{2+} + 3H_2O + (SCN)_2$$

$$MnO_2 + 4H^+ + 2Cl^- \rightarrow Mn^{2+} + 2H_2O + Cl_2$$

The above points clearly confirm a close similarity between halogens and pseudo-halogens.

HALOGENS AND PSEUDO-HALOGENS

Pseudohalogens undergo polymerisation and form polymerised species. For example:

$$n(CN)_2 \xrightarrow{500^\circ C} 2(CN)_n$$

$$n(SCN)_2 \xrightarrow{\text{Room temp.}} 2(SCN)_n$$

$$n(SeCN)_2 \xrightarrow{\text{On standing}} 2(SeCN)_n$$

Halogens are having no tendency to undergo polymerisation.

Oxidising Power of Halogens and Pseudohalogens

The oxidising power of halogens and pseudohalogens gets decreased in the following order : $F_2 > Cl_2 > Br_2 > (CN)_2\ (SCN)_2 > I_2 > (SCSN_3)_2 > (SeCN)_2$

This order of oxidising power has been also supported by the standard reduction potential values (E° values) for the following reduction half-reactions :

$$1/2\ Br_2 + e^- \rightarrow Br^- \ ;\ E^\circ = +\ 1.07\ V$$

$$1/2(SCN)_2 + e^- \rightarrow SCN^- \ ;\ E^\circ = +\ 0.77\ V$$

$$1/2\ I_2 + e^- \rightarrow I^- \ ;\ E^\circ = +\ 0.535$$

$$1/2(SCSN_3)_2 + e^- \rightarrow SCSN_3^- \ ;\ E^\circ = +\ 0.275\ V$$

These values reveal that with the decrease of E° values from Br_2 to $(SCSN_3)_2$, the oxidising power of these species also gets decreased in the same direction. Thus :

(a) Br_2 is able to oxidise SCN^- ions to $(SCN)^2$

$$Br_2 + 2SCN^- \rightarrow (SCN)^2 + 2Br^-$$

(b) $(SCN)_2$ is able to oxidise I^- ions to I_2

$$(SCN)_2 + 2I^- \rightarrow 2SCN^- + I_2$$

(c) I_2 is able to oxidise $SeCN^-$ ions to (SeCN)2

$$I_2 + 2SeCN^- \rightarrow (SeCN)_2 + 2I^-$$

HALIDE IONS AND PSEUDOHALIDE IONS

(i) Formation of ionic as well as covalent compounds. Like halide ions, pseudohalide ions are also able to form ionic as well as covalent compounds. Examples of ionic compounds formed by halide ions are AgCl, PbI_2, etc. and the analogous ionic compounds given by pseudohalide ions are AgCN, $Pb(CNS)_2$ etc.

Similarly the examples of covalent compounds formed by halide ions have been ICl, $SiCl_4$, $COCl_2$, SO_2Cl_2 etc. and the analogous covalent compounds given by pseudohalide ions have been ICN, $Si(CN)_2$, $CO(N_3)_2$, $SO_2(N_3)_2$.

(ii) Oxidation of hydracids. HX acids ($X = Cl^-, Br^-, I^-$) can be oxidised to free halogen (X_2). Similarly HY acids ($Y = SCN^-$ CN^- etc.) can also get oxidised to free pseudohalogen (Y_2).

For example.

$$4HCl + MnO_2 \rightarrow MnCl_2 + 2H_2O + Cl_2$$

$$4HSCN + MnO_2 \rightarrow Mn(SCN)_2 + 2H_2O + (SCN)_2$$

(iii) Thermal decomposition of Pb(IV) salts. When heated some of the Pb(IV) salts of halide ions (*e.g.*, $PbCl_4$) decompose to yield Pb(II) salts and free halogen molecule. Similarly some Pb (IV)salts of pseudohalide ions [*e.g.*, $Pb(SCN)_4$] decompose to yield Pb(II) salts and free pseudohalogens. For example :

$$PbCl_4 \xrightarrow{\Delta} PbCl_2 + Cl_2\ ;$$

$$Pb(SCN)_4 \xrightarrow{\Delta} Pb(SCN)_2 + (SCN)_2$$

(iv) Formation of complex ions. Similar to halide ions, pseudohalide ions are also known to form complex ions with transition metal ions. Examples of complex ions formed by halide ions with transition metals have been $[FeF_6]^{3-}$, $[CoCl_4]^{2-}$, $[HgI_4]^{2-}$ etc. and the analogous complex ions given by pseudohalide ions are $[Fe(CN)_6]^{3-}$, $[Co(SCN)_4]^{2-}$, $[Hg(CN)_4]^{2-}$ etc.

(v) Formation of inter-pseudohalogen and inter-halogen compounds. Halide ions can combine together to form interhalogen compounds such as ClF, ICl, IBr etc. Similarly pseudohalide ions can also combine together to form inter-pseudohalogen compounds such as $CN.N_3$, CN.SCN etc.

(iv) Formation of monobasic hydracids. Similar to halide ions, pseudohalide ions also combine with H_2 to form monobasic hydracids.

$$H_2 + 2Cl^- \rightarrow 2HCl + 2e^-$$

$$H_2 + 2CN^- \rightarrow 2HCN + 2e^-$$

(viii) Formation of insoluble salts. Similar to halide ions, pseudo halide ions also yield insoluble salts with Ag^+, Pb^{2+} and Hg^+ cations. Examples of insoluble salts formed by halide ions include AgCl, $PbCl_2$, HgCl etc. and the analogous insoluble compounds given by pseudohalide ions are AgCNS $Pb(CNS)_2$, HgCN, etc.

(ix) Acting as bridging ligands. Similar to halide ions, pseudohalide ions all also having more than one pair of electrons and hence can coordinate with two metal ions simultaneously.

Consequently, like halide ions, pseudohalide ions also can act as bridging ligands. For example in $[R_2AuCl]_2$ which is a dimeric compound, two Cl^- ions act as bridging ligands. Similarly, in $[R_2AuCl]_2$ which is a dimeric compound, two Cl^- ions act as bridging ligands.

$[R_2Au(CN)]_4$ Compound	$[Ru_2AuCl]_2$ Compound
(Tetrametric)	(Dimeric)

Note that in each of the above compounds the geometry of Au atoms is square planar.

DISSIMILARITIES BETWEEN HALIDE IONS AND PSEUDOHAIDE IONS

(i) The hydracids formed by the pseudohalide ions on combination with hydrogen have been relatively weaker than the analogous hydracids formed by the halide ions. For example HCN is a very weak acid whereas HCl is a very strong acid in aqueous medium. The weak nature of hydracids given by pseudohalide ions may be due to the poor electronegativity of the pseudohalogens.

(ii) Pseudohalide ions have been found to be stronger coordinating ligands than the halide ions and hence the complexes formed by pseudohyalide ions are low spin (*i.e.,* inner-orbital involving d^2sp^3 hybridisation) while those formed by halide ions are high spin (outer-orbital involving sp^3d^2 hybridisation). For example

$[Fe(CN)_6]^{3-}$ is a low spin complex whereas $[FeF_6]^{3-}$ is a high spin complex.

That the pseudohalide ions are stronger ligands than the halide ions is ascribed to the ability of the electron deficient pseudohalides to form σ bond (from the pseudohalide to the metal) as well as π-bond (from the metal to the pseudohalide).

(iii) Pseudohalide ions, being made up of two hetero atoms, can function as ambidentate ligands. For example class a metals are able to form isothiocyanate complexes like $[Co(NCS)_4]^{2-}$ in which N atoms gets coordinated to Co^{2+} ion while class b metals are able to form thiocyanate complexes like $[Pb(SCN)_4]^{2-}$ in which S atom gets bonded to Pd^{2+} ion. Halide ions are having no tendency to act as ambidentate ligands.

FREE PSEUDOHALOGENS

These may be discussed as individuals.

1. *Cyanogen C_2N_2 or $(CN)_2$:* Probably the best-characterised free halogenoid is cyanogen. This compound was discovered by Gay-Lussac in 1815.

Preparation

(i) Cyanogen may be liberated by heating the cyanides of inactive metals such as silver, mercury or gold in a hard glass or steel tube to dull redness.

$$2AgCN \rightarrow 2Ag + (CN)_2$$

$$Hg(CN)_2 \rightarrow Hg + (CN)_2$$

The cyanogen is collected over water or mercury.

The yield of $(CN)_2$ obtained in this way is not quantitative, since $(CN)_2$ formed polymerises (at 500°) to insoluble paracyanogen, $(CN)_x$.

$$x(CN)_2 \xrightarrow[500^\circ C]{\text{Polymerisation}} 2(CN)_x$$

$(CN)_2$ is set free at a lower temperature and is obtained in better yield by heating a mixture of $Hg(CN)_2$ and $HgCl_2$.

$$Hg(CN)_2 + HgI_2 \rightarrow Hg_2Cl_2 + (CN)_2$$

(ii) It is also liberated by the oxidation of cyanide ion in aqueous solution by Cu^{2+} ion as

$$2Cu^{2+} + 6CN^- \rightarrow 2[Cu(CN)_2]^- + (CN)_2$$

or it may be prepared by the electrolysis of fused cyanides.

$$2CN^- \rightarrow (CN)^2 + 2e^-$$

(iii) An impure sample of $(CN)_2$ is prepared by dropping concentrated KCN solution into a warm solution of copper sulphate.

$$CuSO_4 + 2KCN \rightarrow Cu(CN)_2 + K_2SO_4$$

$$2Cu(CN)_2 \rightarrow 2CuCN + C_2N_2$$

The cyanogen from CuCN may be liberated by warming the solution with ferric chloride.

$$2CuCN + 2FeCl_3 \rightarrow Cu_2Cl_2 + 2FeCl_2 + (CN)_2$$

The cyanogen so obtained contains impurities of carbon dioxide and hydrocyanic acid. However, HCN is removed by passing the gas over cotton wool soaked in silver salt solution.

(iv) $(CN)_2$ can also be prepared by heating ammonium oxalate with P_2O_5.

$$\begin{array}{c} COONH_4 \\ | \\ COONH_4 \end{array} \xrightarrow[(-2H_2O)]{\text{Heat with } P_2O_5} \begin{array}{c} CONH_2 \\ | \\ CONH_2 \end{array} \xrightarrow{-2H_2O} (CN)_2$$

Ammonium oxalate — Oxamide

Properties

(i) It is a colourless, poisonous gas with the odour of bitter almonds. M.P. = – 28°C and B. P. = – 21°C.

(ii) *Burning in air* : It burns in air with a violet coloured flame to yield CO_2 and N_2. During this reaction explosion takes place.

$$(CN)_2 + 2O_2 \rightarrow 2CO_2 + N_2$$

(iii) *Polymerisation* : Although, pure $(CN)_2$ is stable, the pure gas undergoes polymerisation on heating to 500° to yield an insoluble solid, paracyanogen, $(CN)_n$ which, at 800°, regenerates $(CN)_2$ but decomposes above this temperature.

(iv) *Hydrolysis :* It is soluble in water and undergoes slow hydrolysis to yield a number of products, the main products being ammonium formate, urea, oxamide and ammonium oxalate.

(v) Action of alkalies. It dissolves in alkalies (*e.g.*, KOH) thereby forming the cyanide (KCN) and oxy-cyanide (KCNO).

$$(CN)_2 + 2KOH \rightarrow KCN + KCNO + H_2O$$

(vi) Action of alkali metals. It reacts directly with alkali metals to form their cyanides.

$$2K + (CN)_2 \rightarrow 2KCN$$

(vii) *Addition reactions :* Many reactions of $(CN)_2$ involve addition across one or both of the multiple bonds in its molecule. Thus H_2S yields thiocyanaformide and dithio-oxamide while aqueous NH_2OH yields oxaldihydroxamide.

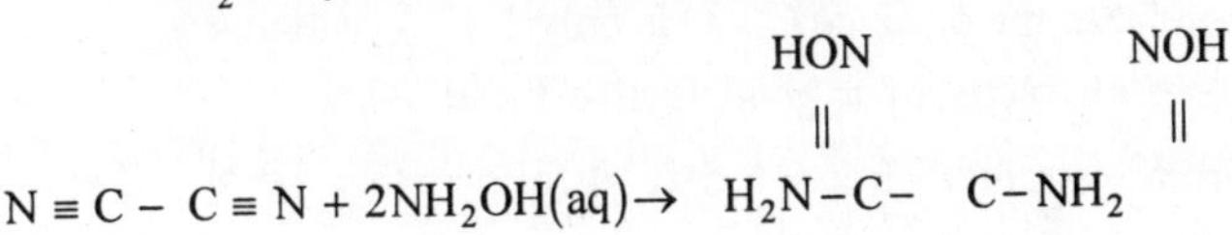

Oxaldihydroxamide.

(viii) *Reduction by hydrogen :* When $(CN)_2$ gets reduced by H_2, both the triple bonds get converted into single bonds by taking up hydrogen atoms and thus ethylenediamine. $H_2N - CH_2 - NH_2$ is formed.

$$N \equiv C - C \equiv N + 4H_2 \quad H_2N - CH_2 - CH_2 - NH_2$$

(ix) *Fluorination :* When $(CN)_2$ is fluorinated by AgF_2 at – 110°, it yields $(F_2CN)_2$ (l). although, other products may also be obtained.

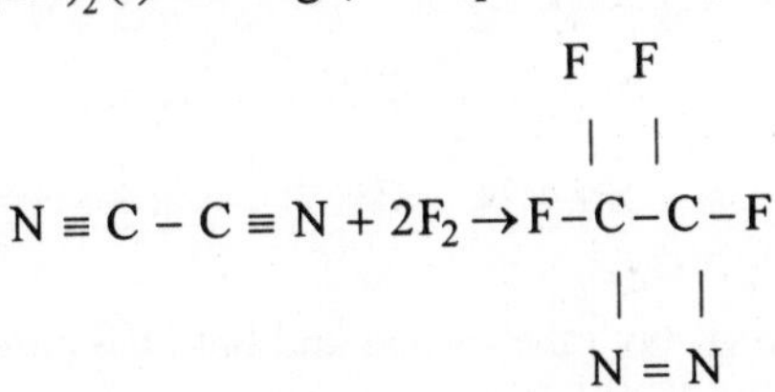

(1)

(x) *Other reactions :* Some other reactions of $(CN)_2$ are given as follows :

Uses

(i) It is used as poisonous gas in wars.

(ii) It is used inorganic synthesis.

Structure

(i) The density measurements of the substance correspond to the formula C_2N_2.

(ii) When it is reduced, it forms diethylene diamine $H_2N - CH_2 - CH_2 - NH_2$. It indicates that the molecule is apparently linear and the order of atoms must be NCCN. This is fully confirmed by physical studies.

(iii) The C – N bond distances is 1.16 Å which is normal for triple bond, but the distance C – C is only 1.37 Å which suggests that some character of π bond is also there.

(iv) From electron diffraction data, it follows that the N – C and C – C bond lengths which are equal to 1.16 Å respectively show; that the true structure of the molecule is best represented as a resonance hybrid of the following three structures, structure (I) being the most contributing form.

(I) :N: ::C:C: : :N: ° C – C ° N)

(II) :N: :C: :N: (N = C = C = N)

(III) $:\overline{N}:\ :C:\ :C:\ \overset{+}{N}:\left(\overline{N} = C = C = \overset{+}{N}\right)$

(IV) $:\overset{+}{N}:\ :C:\ :C:\ :\overline{N}:\ \left(\overset{+}{N} = C = C = \overline{N}\right)$

The III and IV structures show resonance.

$$:N \equiv C - C \equiv N: \leftrightarrow :\overset{+}{N} = C = C = \overline{N}: \leftrightarrow \overline{N} = C = C = \overset{+}{N}:$$

(I) (II) (III)

In structure (I) both the carbon and both the nitrogen atoms are sp hybridised and each carbon and nitrogen pair are able to form two π-bond in between them. One of the two sp-hybrids on each N atom in having a lone pair of electrons and other sp-hybrid orbital has one electron. This singly-occupied sp-hybrid on each N atom overlaps with sp-hybrid on carbon atom to form sp-sp N – C σ–bond. sp hybrid orbitals on each carbon atom overlap to form sp- C – C σ–bond. Thus in all three are, one C – C sp-sp σ–bond, two N–C sp-sp σ–bond and four N–C p-p π–bonds. Each N atom has a lone pair of electrons residing in one of the two sp-hybrid orbitals.

However, the structure of $(CN)_x$ is as follows :

2. Oxycyanogen $(OCN)_2$

Preparation

Evidence for the existence of this compound is marginal but Lidov claimed to have obtained this compound by the oxidation of potassium cyanate with H_2O_2.

$$4KOCN + 2H_2O_2 \rightarrow 2K_2CNO_2 + 2H_2O + (CNO)_2$$

It is also prepared by the action of AgOCN on I_2 in CS_2 or CCl_4 solution.

$$2AgOCN + I_2 \rightarrow 2AgI + (OCN)_2$$

Another method is by the action of silver oxide on cyanogen bromide.

$$Ag_2O + 2CNBr \rightarrow 2AgBr + (OCN)_2$$

This method is also due to Lidov. However, the observation of Lidov has not been confirmed. Birchenbach and Kellermann were unable to isolate the compound as a result of electrolytic oxidation of potassium cyanate in method but they obtained a solution which liberated iodine from iodide and dissolved copper, iron and zinc without the evolution of gases. This is an indirect evidence that oxycyanogen was produced in solution.

At present oxycyanogen needs further investigations.

Properties

$(CNO)_2$ shows typical properties of pseudohalogens. It liberates I_2 from an iodide solution, reacts with metal to forms inter-pseudohalogen compounds.

3. Thiocyanogen : $(SCN)_2$

Preparation

(i) It was first of all prepared by Soderback (1919) by the action of iodine or bromine on a dry ethereal suspension of silver thiocyanate.

$$2Ag(SCN) + Br_2 \rightarrow 2AgBr + (SCN)_2$$

$$2Ag(SCN) + I_2 \rightarrow 2AgI + (SCN)_2$$

(ii) Thoicyanogen may also be prepared by the oxidation of thiocyanic acid with manganese dioxide.

$$2HCNS + [O] \rightarrow H_2O + (SCN)_2$$

(iii) When lead tetra acetate in ether is treated with thiocyanic acid thiocyanogen is formed.

$$(CH3COO)_4Pb + 4HCNS \rightarrow Pb(SCN)_4 + 4CH_3COOH$$

$$Pb(SCN)_4 \rightarrow Pb(SCN)_2 + (SCN)_2$$

$Pb(SCN)_2$ on oxidation yields thiocyanogen.

$$Pb(SCN)_2 + O_2 \rightarrow PbO_2 + (SCN)_2$$

(iv) It is also obtained by treating lead thiocyanate $Pb(SCN)_2$ with Br_2 in ethereal solution at 0°C.

$$Pb(SCN)_2 + Br_2 \rightarrow (SCN)_2 + PbBr_2$$

Evaporation of the solvent after filtering $PbBr_2$ gives free $(SCN)_2$.

(v) It is also obtained by electrolysing an alcoholic solution of NH_4SCN or KSCN.

In all the above methods, thiocyanogen is obtained in ether layer. If this solution is cooled, a red-brick isomorphous solid is formed which readily polymerises to form $(SCN)_x$.

Properties

(i) It melts at –3°C to a yellow oil which polymerises to form the brick-red parathiocyanogen, $(SCN)_x$. It is readily hydrolysed to form thiocyanic acid and hydrothicoyanic acid.

$$(SCN)_2 + H_2O \rightarrow HSCN + HOSCN$$

(iii) With unsaturated hydrocarbons, it forms addition compounds.

$$\begin{matrix} CH_2 & & & CH_2SCN \\ \| & + (SCN)_2 & \rightarrow & | \\ CH_2 & & & CH_2SCN \end{matrix}$$

(iv) Following reactions of $(SCN)_2$ indicate its oxidising property and similarities with halogens.

(a) A solution of $(SCN)_2$ in ether or in CS_2 liberates I_2 from an iodide.

$$(SCN)_2 + 2I^- \rightarrow 2SCN^- + I_2$$

(b) It oxidise Cu(I) salts to Cu(II) salts.

$$(SCN)_2 + 2CuSCN \rightarrow 2Cu^{2+} + 4SCN^-$$

(c) It reacts with metals to form metallic salts.

$$(SCN)_2 + M\ (M = 0) \rightarrow M(SCN)_2\ (M = +2)$$

(d) It oxidises thiosulphite ($S_2O^2_3$–) into tetrathionate ($S_4O^2_6{}^-$)

$$(SCN)_2 + 2S_2O3^{2-} \rightarrow 2SCN^- + S_4O^2_6{}^-$$

(e) Sulphur gets precipitated from H_2S.

$$(SCN)_2 + H_2S \rightarrow 2H^+ + 2SCN^- + S\downarrow$$

(f) Arsenite ($AsO^3_3{}^-$) gets oxidised to arsenate ($AsO^3_4{}^-$).

$$SCN)_2 + AsO^3_3{}^- + H_2O \rightarrow 2SCN^- + 2H^+ + AsO^3_4{}^-$$

(v) Some other reactions shown by $(SCN)_2$ can be represented by the following chemical equations.

(a) Cold solution of $SCN)_2$ + NO $\rightarrow$ $\underset{\text{(Nitrosyl thioyanate)}}{\text{NO.SCN}}$

(b) IM cole solution

$$(SCN)_2 \xrightarrow{+2HCl} (SCN)_2.2HCl \text{ (Dihydro chloride)}$$

$$\xrightarrow[(-2HCl)]{+H_2O} (SCN)_2.H_2O \text{ (Monohydrate).}$$

Uses

(i) It is used as a titrimetic oxidising agent.

(ii) It is used for the determination of unsaturation in organic compounds.

Structure of Thiocyanogen

(i) Cryoscopic measurements give molecular weight corresponding to the molecular formula $(SCN)_2$.

(ii) It has been proposed that there exists an equilibrium between structures I and II.

Evidence in Favour of Structure I

(i) X-ray data of Strada favour the structure I.

(ii) All the chemical properties of $(SCN)_2$ can be explained by structure I.

(iii) A large number of thiocyanate complexes are known in which the SCN^- ion co-ordinates through sulphur or through nitrogen confirming the structure I.

Evidence in Favour of Structure II

The structure II does not possess any experimental evidence.

Thus, the structure I is the most parable structure.

Electronically structure I is represented as

:S: :C: :N:

:S: :C: :N:

Sturcture of SCN^-

It has been shown definitely to have the structure SCN^- but the question of its linearity is not definitely settled.

Structure of $(SCN)_2$

This apparently exists in the dimeric state of certain solvents, but in the free state it rapidly and irreversibly polymetrises to brick-red paracyanogen $(SCN)_x$, the structure of which is not known.

4. Selenocyanogen, $(SeCN)_2$

Preparation

(i) Brickenbach and Kellermann prepared this compound by the electrolysis of a solution of potassium selenocyanate in methyl alcohol. This method is not suitable for the preparation of pure compound as selenium is deposited very readily.

(ii) A pure sample can be obtained by treating excess of silver slenocyanate with iodine in ether at temperature below 0°C.

$$2AgSeCN + I_2 \rightarrow (SeCN)_2$$

(iii) Another better procedure involves the reaction of Pb (IV) acetate in chloroform with potassium selenocyanate in anhydrous acetone.

$$Pb(CH_3COO)_4 + 4KSeCN \rightarrow 4CH_3COOK + Pb(SeCN)_4$$

$$Pb(SeCN)_4 \rightarrow Pb(SeCN)_2 + (SeCN)_2$$

Properties

(i) It is a yellow powder which generally converts into a red material.

(ii) It is soluble in chloroform, carbon disulphide, acetic acid, carbon tetrachloride, benzene, etc. but selenium is deposited readily in these solutions.

(iii) It undergoes hydrolysis readily.

$$2(SeCN)_2 + 3H_2O \rightarrow H_2SeO_3 + 3HSeCN + HCN$$

(iv) It acts an oxidising agent of slightly less strength than iodine which is indicated by the following equilibrium :

$$2AgI + (SeCN)_2 \rightleftharpoons 2AgSeCN + I_2$$

(v) When a solution of selenocyanogen in carbon disulphide is refluxed, polymerisation occurs and the crystalline compound, $(SeCN)_x$ separates out on cooling.

Structure

(i) Its molecular weight in benzene corresponds to the composition $(SeCN)_2$.

(ii) Its structure is similar to thiocyanogen.

:Se: :C: :N: Se – C ≡ N

or |

:Se: :C: :N: Se – C ≡ N

Raman spectral studies have shown that $(SeCN)_2$ molecule has a linear structure.

$$N \equiv C - Se - Se - C \equiv N$$

5. Tellurocyanogen, $Te(CN)_2$

In 1964 Greenwood and Sprague reported that appears to be no evidence for the existence of $TeCN^-$ ion but the compound $Te(CN)_2$ exists. It can be prepared by the interaction of a halide with silver cyanide in benzene.

$$TeBr_4 + 3AgCN \rightarrow Te(CN)_2 + 4AgBr + BrCN.$$

However, this compound, needs further investigation.

6. Azide Carbon Disulphide $(SCSN_3)_2$

Preparation

(i) It is obtained by oxidising potassium azide thiocarbonate with H_2O_2 or I_2.

$$2KCSCN_3 + I_2 \rightarrow 2KI + (SCSN_3)_2.$$

(ii) It may also be prepared by the electrolytic oxidation of $KSCSN_3$.

(iii) Its sodium or potassium salt is prepared by action of potassium azide on carbon disulphide at 44°C.

$$KN_3 + CS_2 \rightarrow KSCSN_3$$

$$NaN_3 + CS_2 \rightarrow NaSCSN_3$$

These salts can be converted into $(SCSN_3)_2$ by I_2 or H_2O_2.

Properties

(i) It is a white crystalline solid.

(ii) It dissolves in water to an extent of 2 parts in 1000 at 21°C.

(iii) Chemically, this compound is unstable and decomposes at room temperature

$$(SCSN_3)_2 \rightarrow 2N_2 + 2S + (SCN)_2.$$

(iv) When it is dissolved in aqueous alkali, it reacts according to the equation.

$$(SCSN_3)_2 + 2OH^- \rightarrow SCSN_3 + OSCSN_3^- + H_2O$$

Structure

The structure of $(SCSN_3)_2$ is unknown. However, it may be either I or II.

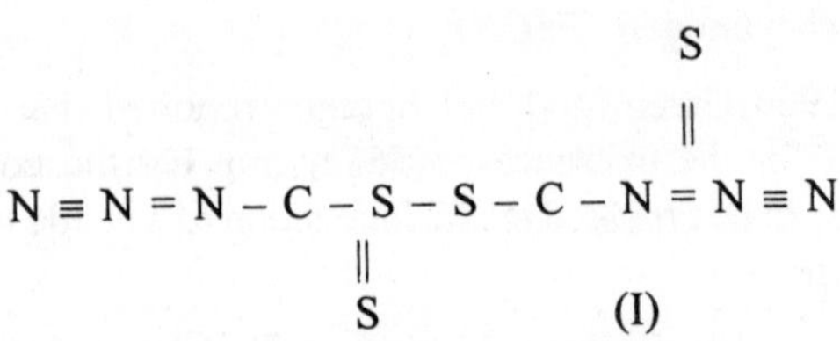

(I)

S S

$$
\begin{array}{c}
\qquad\qquad\qquad \| \qquad\qquad\qquad \| \\
N \equiv N = N - C - S - C - N = N \equiv N \\
\| \\
S
\end{array}
$$

(II)

Probably, the structure (I) is more correct.

7. Azides

The azide radical ($N^{-}3$) or its dimer N_6 is unknown in its free state. In a large number of compounds, it shows resemblance with halogens and pseudohalogens.

The sodium salt is prepared by the following reaction :

$$2NaNH_2(l) + N_2O \xrightarrow{190^{\circ}C} NaN_3 + NaOH + NH_3$$

Free hydrazoic acid (N_3H) is obtained by distilling sodium azide with sulphuric acid (1 : 1)

$$2NaN_3 + H_2SO_4 \rightarrow Na_2SO_4 + 2N_3H$$

Free hydrazoic acid, obtained by distillation from aqueous solution, is colourless volatile liquid. It explodes violently when heated or struck.

Hydrazoic acid dissolves magnesium, zinc, aluminium, etc., to form azides.

$$2N_3H + Mg \rightarrow (N_3)_2Mg + H_2.$$

The structure of the azide ion is linear and symmetrical, and its electronic structure on the basis of valency bond theory may be put as :

$$:\overline{N} \equiv \overset{+}{N} = \overline{N}: \leftrightarrow :N \equiv \overset{+}{N} \overset{2-}{:N} \leftrightarrow \overset{2-}{:N} : \overset{+}{N} \equiv N:$$

In covalent azides, on the other hand, the symmetry is lost as evident in HN3 and CH_3N_3 as given below :

In such covalent azides, the electronic structure is a resonance hybrid of

$$R:N = \overset{+}{N} = \overline{N}: \leftrightarrow R:\overline{N} \overset{+}{N} \equiv N:$$

Uses

(i) Azides are used as organic intermediates.

(ii) Lead azide is used a detonator.

PEROXIDES

Introduction

A peroxide may be regarded as a salt of hydrogen peroxide which contains at least one pair of oxygen atoms bonded by a single covalent bond $(O - O)^{2-}$, and which in addition to its oxidising and reducing properties, behaves as an extremely weak acid.

The O_2^{2-} ion has 18 electrons and the electronic configuration can be represented as $\sigma_1 s^2, \sigma_1{}^* s^2\ \sigma_2 s^2, \sigma_{2*} s^2,$

$$\sigma 2p_x^2, \pi 2p_y^2 = \pi 2p_z^2, \pi^* 2p_z^2 = \pi^* 2p_z^2.$$

This indicates single bond between oxygen atoms (b.0=$\frac{10-8}{2}$ or 1). It also shows diamagnetic nature. It also shows that bond strength of O_2^{2-} ion is less and so bond length is more than heat of O_2 molecule.

Each of the oxygen atoms in peroxides has an oxidation state of minus one (–1) whereas normal oxide contain minus two (–2) oxidation state, while molecular oxygen is considered to be in the zero oxidation state.

Peroxide formation is determined by size and electropositive character of the metal. The ease of formation of peroxides increases with increase in size and electropositive character of the metal.

Peroxides are to be distinguished from several other types of compounds with similar names. The higher oxides of lead, manganese and other, although sometimes incorrectly called peroxides, are not peroxides as defined above because they contain no oxygen-oxygen bond.

In a similar fashion, such compounds as the perchlorates, permanganates perrhenates and perfluoro compounds are not peroxides, in these cases the prefix "per" denotes a fully oxidised state for a central atom.

General Methods for the Preparation of Peroxides

The peroxides may be prepared by the following methods :

1. *Direct reaction of a metal or metal oxide with metal :* Peroxides are obtained by the oxidation of the element or more usually of a lower oxide or hydroxide, with oxygen, hydrogen peroxide or a compound yielding H_2O_2 on hydrolysis.

$$2Na + O_2 \rightarrow Na_2O_2$$

$$P_2O_{10} + 4H_2O_2 + 2H_2O \rightarrow 4H_3PO_5$$

$$N_2O_5 + H_2O_2 \rightarrow HNO_3 + HNO_4$$

When peroxides are prepared by the method mentioned above, they are usually hydrated and contain hydrogen peroxide of crystallisation. However, ozonised oxygen with silver yields silver peroxide. Strontium gives peroxides under 60 atm. pressure.

$$Sr + O_2 \rightarrow SrO_2$$

$$2Ag + O_2 \rightarrow Ag_2O_2$$

2. *Direct reaction of metal dissolved in liquid ammonia with oxygen.* Potassium peroxide, rubidium peroxide, and caesium peroxide cannot be obtained readily by the method described above, These are obtained by oxidising the metal in liquid ammonia solution with oxygen or air or by careful thermal decomposition of the superoxide.

3. *Thermal decomposition of metal superoxides :* This method can be utilised to prepare peroxides of the type M_2O_2 by thermal decomposition of alkali metal superoxides.

4. *Reactions involving hydrogen peroxide :* By this method Na_2O_2, SrO_2, BaO_2 and MgO_2 can be prepared. This method involves the reaction of hydrogen peroxide with metal salt solution in the presence of hydroxyl ions.

Properties of Peroxides : These are outlined as below :

1. *Hydrolysis :* Peroxides reveal their relationship to hydrogen peroxide by forming this compound with solution in water. In some cases, as with sodium peroxide, hydrolysis is essentially instantaneous. In other cases, hydrolysis is very slow in cold solutions.

$$Na_2O_2 + 2H_2O \rightarrow 2NaOH + H_2O_2$$

2. *Stability :* In general, solid inorganic peroxides are much more stable in the dry state than in solution. Some peroxides, as for example sodium peroxide, are thermodynamically stable in the absence of moisture. The stability of metallic peroxides increases with electropositive nature of the metal. Some peroxides are explosive *e.g.,* CuO_2, and HgO_2.

3. *Chemical Reactivity* : The chemical reactivity of peroxides is related to the very low energy of the oxygen-oxygen single bond. In hydrogen peroxide the bond strength is 210 kj/mole which is only about half of a normal bond strength for a single covalent bond.

   The oxidising power of peroxides is due to this low energy coupled with the high energy of oxygen metal, oxygen-carbon and oxygen-hydrogen bonds. The use of peroxides depends largely on their oxidising power. These compounds are very active oxidant.

4. *Oxidising agents* : They are good oxidising agents.
5. They liberate oxygen on heating.

Classification of Peroxides : Metallic peroxides are of two types :

(a) *Ionic Peroxides* : These are formed by the most electronic positive elements and contain O_2^{2-} ion. Ionic peroxides are formed by the alkali and alkaline earth metals. It is almost certain that peroxides of the type MO_2 formed by magnesium, zinc and cadmium are also ionic, though the classification of HgO_2 is less clear.

(b) *Peroxy compounds of the transition elements, lanthanides and actinides* : In these the peroxy group is found in a variety of cationic and anionic species, occasionally as a bridging group, and also as ligand bonded to the metal atom in neutral molecule.

We shall now discuss these one by one by taking different examples.

1. Alkali Metal Peroxides

(a) *Lithium Peroxide, Li_2O_2* : It is a colourless solid, best prepared by precipitation from a solution of lithium hydroxide and hydrogen peroxide followed by careful drying.

$$2LiOH + H_2O_2 \rightarrow Li_2O_2 + 2H_2O.$$

Properties : It is very water soluble, yielding an alkaline hydrogen peroxide solution. No commercial uses of this compound have been developed.

(b) *Sodium Peroxide, Na_2O_2* : It was first prepared commercially be Aluminium Company Limited at oldbury, England, using a process invented by Castner. This company prepared it by burning metallic sodium in a counter current stream of purified and dried air.

$$2Na + O_2 \rightarrow Na_2O_2$$

The modern process involves the following two steps :

(a) *Conversion of sodium to sodium monoxide :* This step is carried out in rotatory steel burners. Sodium in solid or liquid form is fed through the burner shell to bed of sodium and sodium monoxide Dry; air is introduced at one end, and the exhaust gas, substantially nitrogen, and dust pass out through a dust collection chamber. Sodium peroxide is withdrawn periodically.

$$4Na + O_2 \rightarrow 2Na_2O$$

(b) *Conversion of sodium monoxide to sodium peroxide :* This conversion is carried out in a rotating steel finisher. Monoxide is admitted at one end and oxygen at the opposite end. Temperature varies from 240 to 390°C. The sodium peroxide is cooled and is then stored.

$$2Na_2O + O_2 \rightarrow 2Na_2O_2$$

Properties

(i) It is a pale yellow solid which melts without decomposition at about 650°C but evolves oxygen at about 750°C.

(ii) Solid sodium peroxide is a very powerful oxidising agent. All metals including platinum are attacked by the molten compound. Iron and nickel vessels are relatively resistant. It oxidises chromic hydroxide to sodium chromate, manganous salts to sodium manganate and sulphides to sulphates.

$$\left[O_2^{2-} + 2H_2O + 2e \rightarrow 4OH^-\right] \times 3$$

$$\left[Cr(OH)_3 + 3OH^- \rightarrow CrO_4^{2-} + 4H_2O + 3e\right] \times 2$$

$$2Cr(OH)_3 + 3O_2^{2-} \rightarrow 2CrO_4^{2-} + 2OH^- + 2H_2O$$

(iii) It dissolves in water and produces H2O2

$$Na_2O_2 + 2H_2O \rightarrow 2NaOH + H_2O_2$$

(iv) It forms Na_2CO_3 on action with CO and CO_2

$$Na_2O_2 + CO \rightarrow Na_2CO_3 + O_2$$

$$2Na_2O_2 + C \rightarrow 2Na_2CO_3 + O_2$$

(v) It reacts with $COCl_2$ to give sodium percarbonate.

$$2Na_2O_2 + COCl_2 \rightarrow 2NaCl + Na_2CO_4 + 1/2O_2$$

(vi) It reacts with chrome iron and iron pyrites to give Fe_2O_3.

$$2FeO.Cr_2O_3 + 7Na_2O_2 \rightarrow Fe_2O_3 + 4aCrO_4 + 3Na_2O$$

$$2FeS_2 + 15Na_2O_2 \rightarrow Fe_2O_3 + 4Na_2SO_4 + 11Na_2O.$$

Uses

(i) Sodium peroxide in aqueous solution can be used to bleach wood products, paper pulp, straw and cellulosic textiles, such as cotton, linen and jute.

(ii) It can be used in the synthesis of other peroxides, such as sodium peroxyborate, peroxide addition compounds, and various organic peroxides.

(iii) It is the oxidant used in the parr Sulphur bomb procedure for estimating sulphur content of coal, coke and other products.

(c) potassium peroxide K_2O_2, rubidium peroxide, Rb_2O_2 and caesium peroxide Cs_2O_2. These are not readily formed by direct reaction of the elements because of the preferential formation of the superoxide. These peroxides are said to be obtainable by oxidising the metal in liquid ammonia solution with oxygen or air or by careful thermal decomposition of the superoxide.

All the three compounds are hygroscopic and easily subjected to disproportionation reaction. None has practical uses.

2. Silver Peroxide Ag_2O_2

Preparation

(i) It can be prepared by the action of ozonised oxygen on silver metal.

(ii) It can be prepared by anodic oxidation of potassium persulphate.

(iii) It can also be prepared by the action of potassium persulphate on silver nitrate solution.

Properties

(i) It is a dark coloured compound

(ii) It acts as a powerful oxidising agent.

(iii) It is not easily hydrolysed.

3. Alkaline Earth Metal Peroxides

Well defined, stable peroxides of the formula MO_2 are known for the elements such a magnesium, strontium and barium. There is doubt that a true beryllium peroxide has been prepared. All yield hydrogen peroxide when dissolved in water.

(a) Magnesium peroxide, MgO_2

Preparation

(i) The purest magnesium peroxide has been made by the reaction of hydrogen peroxide with alkali and a suitable magnesium salt. Under careful drying conditions, a product approximating $MgO_2.H_2O$ has been obtained.

$$MgSO_4 + H_2O_2 + 2NaOH \rightarrow MgO_2 + Na_2SO_4 + 2H_2O$$

(ii) The magnesium peroxide is made by reacting a suitable grade of light magnesium oxide with strong hydrogen peroxide. Their dried slurry contains upto 50% MgO_2 with balance being mostly MgO.

$$MgO + H_2O_2 \rightarrow MgO_2 + H_2O$$

Properties

(i) It is an off-white solid, similar to magnesium oxide in appearance and physical properties.

(ii) This compound is very slightly soluble in water but dissolves readily in aqueous acid to yield hydrogen peroxide in solution.

$$MgO_2 + 2HNO_3 \text{ aq} \rightarrow Mg(NO_3)_2 + H_2O_2$$

(iii) When heated, this compound decomposes to oxygen gas and normal magnesium oxide.

Uses

This compound finds special uses in pharmacology, as an antifermentative ingredient in certain gastric antacid compositions.

(b) Calcium peroxide, CaO_2

Preparation

It is best made by the reaction of hydrogen peroxide with an alkali, and a soluble calcium salt. The insoluble peroxide can be filtered and dried. The commercial product contains about 60% CaO_2. the balance being mostly calcium oxide.

$$Ca(NO_3)_2 + H_2O_2 + 2NaOH \rightarrow CaO_2 + 2NaNO_3 + 2H_2O$$

Properties

It is an off-white solid with a density of 3.17 g/cm^3. On heating, it dissociates into oxygen and calcium oxide without melting.

$$2CaO_2 \rightarrow 2CaO + O_2$$

Uses

It finds use as a dough conditioner in baking industry.

(c) Strontium Peroxide, SrO_2

Preparation

(i) This can be manufactured by the direct oxidation of strontium cxide at about 200 atm. O_2 pressure at 350°C.

$$2SrO + O_2 \rightarrow SrO_2$$

(ii) It can also be prepared by the action of alkaline hydrogen peroxide on a soluble strontium salt. A dense pure product is obtained if this reaction is carried out above 550°C.

Properties

It is a light coloured powder or granular solid as usually prepared. This compound is thermodynamically stable at room temperature.

Strontium peroxide is a powerful oxidant, when heated resembling the other metal peroxides in this regard and in yielding hydrogen peroxide dissolving in aqueous acid.

Uses

It has found uses in pyrotechnics and in tracer bullet formulations.

(d) Barium Peroxide, BaO_2

Preparation

(i) Barium peroxide can also be prepared by there action of alkaline hydrogen peroxide on a soluble barium salt.

$$BaCl_2 + H_2O_2 + 2KOH \rightarrow BaO_2 + 2KCl + 2H_2O$$

Barium peroxide is made by heating BaO in oxygen at 400°C.

$$BaO + 1/2O_2 \rightarrow BaO_2$$

Properties

It is a ten-coloured product, thermodynamically stable at room temperature and to 500°C and above this, it dissociates to oxygen gas and peroxide.

Barium peroxide displays the reactions typical of the peroxide including powerful oxidising action when heated and formation of hydrogen peroxide on dissolving in aqueous acid.

Uses

It has been used in pyrotechnics but has had no large scale use since hydrogen peroxide production via, BaO_2 ceased in 1940.

4. Peroxides of IIB Group

Zinc and cadmium form similar peroxides like alkaline earth metals peroxides, but mercury, remaining member of group IIB, forms a peroxide which has a transitory existence.

(a) Zinc peroxide, ZnO_2

Preparation

(i) Zinc peroxide containing about 80 per cent of ZnO_2 is formed in the reaction of ethereal hydrogen peroxide with diethyl zinc.

$$2Zn(C_2H_5)_2 + 3H_2O \rightarrow 4C_2H_6 + 2ZnO_2 + H_2O + 1/2O_2$$

(ii) It is manufactured by drying a slurry of zinc oxide in strong hydrogen peroxide.

$$ZnO + H_2O_2 \rightarrow ZnO_2 + H_2O$$

(iii) Anhydrous ZnO_2 is prepared by adding 0.15 M NaOH to $ZnSO_4$ dissolved in 30% hydrogen peroxide and drying precipitate over P_2O_5.

Properties

It is very similar to magnesium peroxide in both appearance and behaviour. It decomposes at the ordinary temperature and gives H_2O_2 what water.

Uses

The chimerical product containing 55% ZnO_2 is used in dusting powders, as a deodorant and in pharmacology for the treatment of infections and skin lesions. It is useful in certain rubber formulations, combining the functions of filter and pigment with cross-linking action. It is used as a deodorant.

(b) Cadmium Peroxide, CdO_2

Preparation

It is made by the action of 30% H_2O_2 on an ammoniacal solution of $CdSO_4$ and drying the precipitate product at 120°C.

$$CdSO_4 + 2NH_4OH + H_2O_2 \rightarrow CdO_2 + (NH_4)_2\,SO_4 + 2H_2O$$

It may also be obtained by the reaction of aqueous hydrogen peroxide a slurry of the oxide.

$$CdO + H_2O_2 \rightarrow CdO_2 + 2O$$

Only impure products of cadmium peroxide have been obtained and these are insufficiently stable to be practical interest.

(c) Mercuric Oxide, HgO_2

It exists in two crystalline forms, It is produced by the action of concentrated hydrogen peroxide solution on mercury (II) oxide.

$$HgO + H_2O_2 \rightarrow HgO_2 + H_2O$$

It is unstable and evolves oxygen on treatment with water. It explodes on striking or on rapid heating.

Structure of Second Group Peroxides

The ionic radii of magnesium, zinc and cadmium are too small in relation to that of O_2^{2-} for the calcium carbide structure to be adopted, and these three compounds all have pyrite's structure.

Only the structure of one form of HgO_2 has been determined. The symmetry is orthorhombic, with the mercury atoms and peroxide groups forming infinite zig-zag chains extending along the O-axis.

This structure is probably best thought of as involving covalent bonding between the metal and oxygen and it is likely that in ZnO_2 and CdO_2 the bond is also largely covalent.

5. Peroxide of Titanium

When a titanium (IV) solution is treated with hydrogen peroxide under approximately neutral conditions, addition of alcohol leads to precipitation of a yellow solid, $TiO_3.2H_2O$.

This is found to have an oxidising capacity, as measured by the liberation of iodine from an iodide, corresponding with the presence in the molecule of one peroxy group.

Some Properties of $TiO_2.2H_2O$ are as Follows

(i) It is a yellow solid.

(ii) It is a powerful oxidizing agent and liberates iodine from iodide.

(iii) It reacts with alkali and H_2O to form $M_4\ Ti(O_2)_4$ type of peroxytitanates.

(iv) It get gydrolysed to form $Ti(OH)_3OOH$.

(v) It reacts with hydrogen peroxide in the presence of sulphuric acid to form $Ti(O_2)SO_4$.

Addition of alkali and H_2O_2 to the above compound *i.e.*, $TiO_3\ 2H_2O$ yields a peroxytitanate of the type $M_4Ti(O_2)_4$. The same compound is formed when a titanium (IV) solution is treated with H_2O_2 and alkali, anhydrites of the type $M_4Ti(O_2)_4.\ 6H_2O$ may be isolated.

The titanium salt decomposes readily at room temperature, with loss of oxygen, and also hydrolyses in solution to $Ti(OH)_3OOH$. Since analysis shows the presence of four peroxy groups per molecule, formulation of the anion as $\left\{Ti\left\{\begin{matrix}O\\|\\O\end{matrix}\right\}_4\right\}^{4-}$ is indicated, with each peroxy ligand contributing a charge of –2 to the complex.

Addition of H_2O_2 to an acidified titanium (IV) solution gives an orange red colour which is used as a sensitive analytical test for titanium and for H_2O_2. The nature of the compound responsible for this colour has not been established, but it appears to contain one peroxy group per metal atom. Thus, in sulphuric acid solution, the reaction with H_2O_2 would be

$$TiO(SO_4) + H_2O \rightleftarrows Ti(O_2)SO_4 + H_2O$$

The solid compound $Ti(O_2)SO_4\ 3H_2O$ has not been isolated.

6. Peroxide of Zirconium

Reaction of a Zr (IV) salt in acid or alkali solution with H_2O_2 produced a white precipitate which has been formulate as $ZrO_3.\ nH_2O_2$ where n varies between 3 and 5. These compounds contain one peroxy group per zirconium atom.

The reaction of Zr (IV) salts with an excess of H_2O_2 results in a peroxide hydrolysis; the final product is represented by the formula ZrO_3, $2H_2O$ or perhaps $Zr(OH)_3\ O_2H$. The possible reactions are described as follows :

(i) $Zr(IV)\text{ salt} + 4H_2O_2 \rightarrow Zr(O_2H)_4 \xrightarrow{3H_2O} Zr(OH)_3(O_2H) + 3H_2O_2$

(ii) $Zr(OH)_4 + H_2O_2 \rightarrow Zr(OH)_2\ (O_2H) + H_2O$

(iii) $ZrO_22H_2O + H_2O_2 \rightarrow ZrO_32H_2O$

(iv) $ZrOH)_4 + O \rightarrow Zr(OH)_3O_2H$

The application of phases diagram and thermogravimetric analysis to the between H_2O_2 and zirconium (IV) hydroxide in the range 0°C to –29°C has shown that the primary product is $ZrO_2.2H_2O$, which may than undergo successive dehydration to give ZrO_3 though anhydrous compound was not isolated.

The peroxy zirconium hydrate is said to be obtained by the reaction of hydrated zirconia ($ZrO_2.2H_2O$) with NaOCl. The reaction of $ZrO_3.2H_2O$ with ice-cold alkali metal hydroxide in the presence of excess H_2O_2 produces a white solid, which contains the teraperoxy zirconide ion, $[Zr(O_2)_4]^{2-}$; a close physical study of the production of the reaction between Zr (IV) and alkaline H_2O_2 has shown that the appearance and composition of the product depends upon the condition of its formation.

The reaction of H_2O_2 with ice cold aqueous solution of zirconium (IV) sulphate produces a white solid of formula $Zr_2O_6SO_4.8H_2O$. It is suggested that this is a basic peroxosulphate structure but there is no physical evidence to support this.

$$\left\{(O_2)Zr \overset{(O_2)}{\underset{SO_4}{\quad}} Zr(O_2)\right\}.8H_2$$

The formation of this compound has been proposed as a method of separating zicronium from hafnium. A complex peroxosulphato-zirconate, analogous to that of titanium is formed when H_2O_2 is added to an ice-cole mixture of Zr (IV) K_2SO_4. and H_2SO_4.

Spectrophotometric studies have shown that zirconium does not form a peroxy species in cone. acid solution. The product obtained by refluxing (ZrO_2Cl_2) with urea and H_2SO_4 of composition H_3ZrO_4. has found some application as a stationary phase oxidant in chromatography.

7. Peroxides of Vanadium

Well defined tetraperoxy compounds the $[V(O_2)_4]^{3-}$ anion are obtained from V_2O_5, alkali hydroxide, and H_2O_2 at 0°C. The solutions are blue and the salts, *e.g.*, $[K_3V(O_2)_4]$ are precipitated by adding ethanol. The potassium salt is vanadium (V) compound and is isomorphous with $K_3[Cr(O_2)_4]$.

When H_2O_2 is added to a solution of vanadyl sulphate containing H_2SO_4, a red colour is produced which is believed to be due to the formation of peroxyvanadyl cation. If excess of H_2O_2 is taken the red colour changes to yellow due to the conversion of peroxyvanadyl cation into $[VO_2(O-O)_2aq]_2$ ion.

This ion is stable in weakly acidic or in alkaline solution and contains two peroxy groups per atom of vanadium. Hence it is an anion of diperoxy orthovanadic acid,

$$[VO.aq]^{3+} + H_2O_2 \rightleftarrows \underset{\text{Red colour}}{[VP_2aq]^{3+}} + H_2O$$

$$[VO.aq]^{3+} + H_2O_2\,2H_2O \quad \underset{\text{Yellow colour}}{\left[VO_2(O-O)_2\right]^{3-}} + 6H^+$$

8. Peroxides of Niobium and Tantalum

For niobium and tantalum, alkali metal salts containing the $[M(O_2)_4]^{3-}$ have been prepared by adding ethanol to a solution containing niobium or tantalum pentoxide, hydrogen peroxide and an alkali metal hydroxide.

These are reasonably stable and dissolve in water without decomposition. Acid converts them into salts containing the $[M(O_2)O_2]^-$ ion.

$$[M(O_2)_2]^{3-} + 2H_2O + 2H^+ \rightarrow [M(O)_2]^- + 3H_2O_2$$

9. Chromium Peroxides

Chromium is known to form two peroxides such as chromium pentoxide and chromium tetroxide which are discussed one by one.

Chromium Pentoxide, CrO_5

Preparation

(i) When hydrogen peroxide reacts with acidic solution of potassium dichromate, a blue solution is obtained which gets discharged after some time. It is believed that the blue colour is attributed to the chromium pentoxide which could not be isolated.

(ii) When 97% H_2O_2 reacts with solution of chromium trioxide in methyl ether an –80°C dark blue solution is obtained. If the excess of CrO_3 is removed by evaporation at –30°C very explosive dark blue crystals, $(CH_3)_2O \rightarrow CrO_5$, get formed which is stable at –50°C.

(iii) It can be obtained in the form of $CrO_5\ H_2O.\ 2NH_3$, when H_2O_2 reacts with CrO_3 in anhydrous ethyl acetate solution followed by alcoholic ammonia.

Properties

(i) It is unstable in aqueous solution but stable in ether for sometime.

(ii) It reacts with dilute acids to yield chromic salts.

$$2CrO_5 + 3H_2SO_4 \rightarrow Cr(SO_4)_3 + 3H_2O + 7O$$

(iii) It reacts with alkali oxides and silver oxide to yield corresponding chromiate.

$$K_2O + CrO_5 \rightarrow K_2CrO_4 + O_2$$

$$Ag_2O + CrO_5 \rightarrow Ag_2CrO_4 + O_2$$

(iv) it yields chromic acid with acidic potassium permanganate.

$$4KMnO4 + CrO_5 + 6H_2SO_4 \rightarrow 2K_2SO_4 + 4MnSO_4 + 10O_2 + 5H_2CrO_4 + H_2O$$

(v) In ether solution CrO_5 reacts with alcoholic H_2O_2 in alkaline medium to form $K_2Cr_2O_{12}.\ 2H_2O$ (blue) and K_3CrO_8 (red) salts.

$$2CrO_5 + H_2O_2 + 2KOH \rightarrow K_2Cr_2O_{12}\ 2H_2O \text{ (blue)}$$

$$2CrO_5 + 3H_2O_2 + 6KOH \rightarrow 2K_3CrO_8\ 6H_2O \text{ (red)}$$

(vi) Its etherial solution reacts with pyridine (C_5H_5n, Pyr) to form a coordination compound, which has been formed to be insoluble in pyridine but soluble in other organic solvents.

The molecule (C_5H_5N) CrO_5 is having the form of a distorted pentagonal pyramid (or trigonal pyramid if O_2 is counted as one ligand) as depicted. The two oxygen atoms of each O_2 group being equidistant *i.e.,* 1.80Å from the metal atom. The O – O bond distance is 1.44Å whereas Cr = O distance is 1.64Å.

(vii) Its reaction with $KMnO_4$ reveals the presence of two peroxy groups in a molecule of CrO_5. It is diamagnetic and is having the following orbital diagram :

Chromium Tetroxide, CrO_4

Preparation

This is known in the form of $(NH_3)_3$. CrO_4 which is a quite stable form. It is prepared by the reaction of solution of $(NH_4)_2$ CrO_4 and H_2O_2 below 0°C and the solution then warmed to 50°C.

Properties

(i) It is a brown coloured solid having the following Structure. It consist of pentagonal bipyramidal molecule in which the four oxygen atoms, one nitrogen and Cr and coplanar. The O–O bond is anomalously short (1.31Å) but the widely different lengths found for this bond in tetaperoxo ion reveal that none of the O–O bond length is very accurately known.

(ii) It reacts with KCN to form K_3CrO_4.3CN and an analogous substance, K_3CrO_4.2CN, NH_3.

(iii) It is a paramagnetic compound which reveals the presence of two unpaired electrons. It is having the following orbital diagram.

Detection and Estimation of Peroxides in Transition Metal Compounds

It is not always easy to prove that particular compound contains peroxide oxygen, or to discover the amount of peroxide oxygen present: the only ultimate criterion is X-ray crystallography. In most cases chemical tests, coupled with elemental analysis are capable of showing conclusively whether peroxide is present or not, but in some compounds, notably those in which other, readily oxidizable groups are present as well as peroxide, chemical methods may not be very reliable. In the chemistry of "main group" peroxides a distinction has been drawn between "bond" (or "true") peroxide and "H_2O_2 adducts" (or "false" peroxides) in transition metal derivatives, the former class includes 'complex' peroxides behave in aqueous solution as do the H_2O_2 adducts.

Attempts have been made to devise chemical tests to distinguish between these two classes; since such a distinction must depend on the resistance of the complex peroxide to hydrolysis, it is found to be to that extent unreliable.

Physical Methods of Determining Peroxy Groups

The only reliable method of determining peroxy groups in solids is that of X-ray crystallography. Few structures of transition metal peroxides have been determined and, of these, few have R values low enough to provide unequivalent proof that the structures are correct in detail (R value is a measure of the agreement between observed and predicted diffraction patterns).

Recent work has shown that the in frated spectra of many transition metal peroxides contain strong absorption in the range 800-900 cm^{-1}, which has been assigned to stretching modes involving the peroxy groups.

Chemical Methods

In a number of simpler transition metal peroxides, the peroxide oxygen has been determined by measuring the amount of O_2 evolved on thermal decomposition. Where there is on other oxidizable group present, and where the transition metal does not change its oxidation state on thermal decomposition of the peroxide, half a molecule of O_2 is evolved to each peroxy group decomposed, the remaining oxygen atom remains as a metal oxygen system. This method is useful and reliable, provided the compound decomposes cleanly and simply.

Acidified $KMnO_4$ solution is generally used in the determination of H_2O_2 and its use in the analysis of transition metal peroxides seems to be accurate and reliable, so long as the necessary precautions are taken, particularly with respect to other redox system, that may be present. Cerium (IV) has been used similarly. Both of these reagents give measure of the peroxide content. Potentiometric titrations, using sodium hypochlorite and potassium ferricyanide, of H_2O_2 liberated from peroxy compounds on hydrolysis in alkaline solution have been used to estimate the amount of 'active' oxygen in the compounds. The use of acetone extraction at 0°C has been suggested, but this method is limited in scope, and has received little attention.

HYDROPEROXIDE

When sodium peroxide is added to ethyl alcohol having a little sodium ethoxide, sodium hydroperoxide is produced.

$$NaO_2 + C_2H_5OH \rightarrow NaOC_2H_5 + NaOOH$$

In the absence of $NaOC_2H_5$ the compound reacts with the alcohol :

$$NaOOH + C_2H_5OH \rightarrow NaOC_2H_5 + H_2O_2$$

Following this further reaction occurs

$$2NaOOH + H_2O_2 \rightarrow (NaOOH)_2H_2O_2$$

A compound in the same class potassium hydroperoxide $(KOOH)_2.3H_2O$ has also been isolated.

PERACIDS AND THEIR SALTS

1. *Definition :* A peroxy acid or peracid is one which contains the peroxide linkage (– O – O) and either it is formed by the action of H_2O_2 on the normal acid or it gives H_2O_2 on treatment with dilute H_2SO_4.

Hence there are two types of peroxyacids (or per acids) namely, the peroxymono acids containing the group – O – O – H, in which one hydrogen atom is replaced by one acidic group and the peroxydiaacids containing the group – O – O, in which both hydrogen atoms are replaced by acidic groups *e.g.*,

$HSO_3 - O - O - H$	$HSO_3 - O - O - SO_3H$
Peroxymono sulphuric acid	Peroxydisulphuric acid

Unfortunately, "per" names for a few acids and salts which are not derivatives of H_2O_2, have become so familiar that their names could not be changed. For this reason, the acids and salts derived from H_2O_2 have been described by the prefix 'peroxy' and not by per-acids.

2. *Test of peracids :* The empirical Riesenfeld test involves the addition of the acid or a salt to a 30% KI solution buffered to pH 7.5 – 8 with $NaHCO_3$. It is claimed that only true peroxoacids oxidiese the

KI to iodine under these conditions. However, the only unambiguous evidence for the peroxide link must come from determinations of structure.

3. *Peroxyacids formed by elements :* The hydrogen peroxide gives rise to two classes of derivatives namely the peroxymono acids containing the group O – O – H, and the per-oxydiacaids containing the group – O – O – in which one or both hydrogen atoms have been replaced by univalent acidic groups respectively. A number of elements form per acids. Out of all these per acids, the per-acids of sulphur, nitrogen, phosphorus and carbon are of importance. Elements which form per-acids are :

Period	*Group 3*	*Group 4*	*Group 5*	*Group 6*
II	b	c	n	
III				S
IV		Ti	V	Cr
V		Zr	Nb	Mo
VI	Ce	Hf,Th	Ta	W

4. General Methods of Preparation of Per-acids

(i) Reactions involving hydrogen peroxide : Many per-acids can be prepared by the action of hydrogen peroxide on acids and their salts.

$$HSO_3Cl + H_2O_2 \rightarrow H_2SO_5 + HCl$$

$$2HSO_3Cl + H_2O_2 \rightarrow H_2S_2O_8 + HCl$$

$$P_4O_{10} + 4H_2O_2 + 2H_2O \rightarrow 4H_3PO_5$$

$$N_2O_5 + H_2O_2 \rightarrow HNO_3 + HNO_4$$

This method is commonly used.

(ii) *Anodic oxidation :* Per acids and their salts can be obtained by the electrolysis of concentrated solution of the acids or their salts by passing a current of high density. This method is used from the preparation of perdisulphuric acid and its salts, potassium peroxydicarbonate and potassium peroxydiphosphate.

(iii) *Reactions involving sodium peroxide :* By this method sodium peroxy-monocarbonate and peroxy-dicarbonate have been prepared.

$$Na_2O_2 + CO_2 \rightarrow Na_2CO_4$$

$$Na_2O_2 + 2CO_2 \rightarrow Na_2C_2O_6$$

$$2Na_2O_2 + COCl_2 \rightarrow Na_2CO_4 + 2NaCl + 1/2O_2$$

(iv) *Reactions involving fluorine :* Per-acids and their salts have been obtained by the action of fluorine on the aqueous solution of the acids or their salts. An example is :

$$2Na_2CO_3 + F_2 \rightarrow Na_2C_2O_6 + 2NaF$$

Properties

(i) Very few peroxyacids have been isolated in the pure form because they are unstable and are often explosive compounds.

(ii) Generally peroxyacids on hydrolysis yield the hydrogen peroxide.

(iii) Peroxy acids do not show reducing properties and do not decolourise permanganate.

5. Per Acids of Sulphur

There are two persulphuric acids, viz,

(i) Permonosulphuric acid or Caro's acid, H_2SO_5

(ii) Perdisulphuric acid or Marshall's $H_2S_2O_8$

These acids are derived from hydrogen peroxide by the replacement of one and two hydrogen atoms by sulphonic groups.

O–H	O–H	O–SO_2OH
|	|	|
O–H	O–SO_2OH	O–SO_2OH
Hydrogen peroxide	Caro's acid	Marshall's acid

(a) Preparation of Marshall's acid $H_2S_2O_8$

(i) From Chlorosulphonic acid : By the action of H_2O_2 (2 mole) on chlorosulphonic acid (1 mole), perdisulphuric acid is obtained.

O – H		Cl SO_2OH		O – SO_2OH	
|	+		→	|	+ 2HCl
O – H		Cl SO_2OH		O – SO_2OH	

(ii) *From H_2SO_4 by electrolysis :* Perdisulphuric acid may be prepared by the electrolysis of 50-60% sulphuric acid at low temperature. The possible mechanism is :

$$H_2SO_4 \Leftrightarrow \underset{\text{Cathod}}{H^+} + \underset{\text{Anode}}{HSO_4^-}$$

At cathode :

$$H^+ + e^- \rightarrow H$$

$$H + H \rightarrow H_2$$

At anode :

$$2HSO_4^- \rightarrow 2HSO_4 + 2e^-$$

$$2HSO_4 \rightarrow H_2S_2O_8$$

Electrolyte : 50 to 60 % H_2SO_4.

Anode : Pt wire surrounded by glass tube serves as a diaphragm.

Cathode : A spiral of copper wire wound round the tube is acting as a cathode. A glass tube containing 0 – 60% H_2SO_4 is placed in a breaker containing ice. In the test tube there is another tube open at both ends which is wound up by a spiral of copper wire that acts as cathode. In this tube, a capillary is kept into which Pt wire is fitted to act as an anode. On passing electric current of about 1 ampere, hydrogen appears at cathode while perdisulphuric acid is formed at anode.

Properties

(a) It is colourless, hydroscopic, crystalline compound (m.p. 65°C.) It is fairly stable at ordinary temperature.

(b) *Action of heat :* It loses oxygen on warming.

$$2H_2S_2O_8 \rightarrow H_2SO_4 + 2SO_3 + O_2$$

(c) *Hydrolysis :* In aqueous solution, perdisulphuric acid is slowly hydrolysed into sulphuric acid and permonosulphuric acid, H_2SO_5.

$$H_2S_2O_8 + H_2O \rightarrow H_2SO_4 + H_2SO_5$$

On distilling with dilute H_2SO_4, H_2O_2 is formed.

$$H_2S_2O_8 + 2H_2O \rightarrow 2H_2SO_4 + H_2O_2$$

(d) *Oxidising character :* It acts as a strong oxidising agent, *i.e.,* it accepts electrons from a reducing agent.

$$2O_8^{2-} + 2e^- \rightleftarrows 2SO_4^{2-}$$

Some examples illustrating its oxidising action are :

(i) It liberates iodine from iodides and bleaches indigo.

$$2KI + H_2S_2O_8 \rightarrow K_2SO_4 + H_2SO_4 + I_2$$

(ii) The salts of the acid are powerful oxidising agents. They oxidise manganese, silver, lead, cobalt and nickel salts to the corresponding metal peroxides.

$$Mn^{2+} + 2H_2O + S_2O_8^{2-} \rightarrow MnO_2 + 4H^+ + 2SO_4^{2-}$$

$$pb^{2+} + 2H_2O + S_2O_8^{2-} \rightarrow PbO_2 + 4H^+ + 2SO_4^{2-}$$

(iii) The salts of the acid liberate chlorine, bromine and iodine from the corresponding halides and even oxidise free iodine to iodic acid.

$$2Cl^- + S_2O_8^{2-} \rightarrow Cl_2 + 2SO_4^{2-}$$

$$2I^- + S_2O_8^{2-} \rightarrow I_2 + 2SO_4^{2-}$$

$$I_2 + 6H_2O + 5S_2O_8^{2-} \rightarrow 2IO_3^- + 10SO_4^{2-} + 12H^+$$

(iv) Salts of perdisulphuric acid oxidise ferrous to ferric, and chromic salts to chromates. However, metallic zinc and copper are oxidised to their sulphates.

$$2Fe^{2+} + S2O_8^{2-} \rightarrow 2Fe^{3+} + 2SO_4^{2-}$$

$$Zn + S2O_8^{2-} \rightarrow Zn^{2+} + 2SO_4^{2-}$$

Uses

(i) For the manufacture of hydrogen peroxide.

(ii) As oxidising agent in dyeing industry.

(b) Permonosulphuric Acid, H_2SO_5

Preparation

(i) From chlorosulophonic acid : It can be prepared by the action of H_2O_2 (1 mole) on chlorosulphonic acid (1 mole).

$$\begin{array}{l} O-H \\ | \\ O-H \end{array} + \xrightarrow{Cl\ SO_2OH} \begin{array}{l} O-SO_2OH \\ \\ O-H \end{array}$$

(ii) From perdisulphuric acid and their salts. In aqueous solution, perdisulphuric acid is slowly oxidised into sulphuric acid and permonosulphuric acid, H_2SO_5.

$$H_2S_2O_8 + H_2O \rightarrow H_2SO_4 + H_2SO_5$$

From $K_2S_2O_8$, it is prepared by keeping a mixture of concentrated sulphuric acid and potassium perdisulphate in an ion bath for an hour and then pouring the mixture over ice.

$$K_2S_2O_8 + H_2SO_4 + \rightarrow K_2SO_4 + H_2S_2O_8$$

$$H_2S_2O_8 + H_2O \rightarrow H_2SO_5 + H_2SO_4$$

(iii) By electrolysis : If concentrated sulphuric acid is submitted to anode oxidation, permonosulphuric acid is produced.

(iv) *From sulphur trioxide* : It may also be prepared by treating sulphur trioxide with 100% H_2O_2.

$$H_2O_2 + SO_3 \rightarrow H_2SO_5$$

Properties

(i) The anhydrous state of permonosulphuric acid is a white crystalline solid (m. p. 45°C). It is stable for some days but slowly loses ozonised oxygen.

(ii) *hydrolysis* : On distilling with dil. H_2SO_4, H_2O_2 is formed.

$$H_2SO_5 + H_2O \rightarrow H_2SO_4 + H_2O_2$$

(iii) Oxidising character :

It acts as a strong oxidising agent that perdisulphuric acid and can liberate chlorine, bromine and iodine from their respective halides instantaneously (distinction from $H_2S_2O_8$).

$$2KI + H_2SO_5 \rightarrow K_2SO_4 + H_2O + I_2$$

$$2KBr + H_2SO_5 \rightarrow K_2SO_4 + H_2O + Br_2$$

It oxidises aniline to aniline black instantaneously. However, like perdisulphuric acid it does not react with chromic acid or postassium permanganate.

Its salts are of no importance.

(c) Structure of Per-acids of sulphur. Permonosulphuric and perdisulphuric acids have been shown to posses the following structures :

S ↑ HO – S– O– OH ↓ O	O O ↑ ↑ HO–S–O–O–S–OH ↓ ↓ O O
Permonosulphuric acid	Perdisulphuric acid

These structures have been confirmed as follows :

(i) By their preparation from chlorosulphonic acid

$$\begin{array}{c} O \\ \uparrow \\ HO - S - Cl \\ \downarrow \\ O \end{array} + H\,O - OH \rightarrow \begin{array}{c} O \\ \uparrow \\ HO - S - O - H \\ \downarrow \\ O \end{array}$$

Permonosulphuric acid

$$\begin{array}{c} O \\ \uparrow \\ HO - S - Cl \\ \downarrow \\ O \end{array} + H + H + \begin{array}{c} O \\ \uparrow \\ Cl - S - OH \\ \downarrow \\ O \end{array} \rightarrow \begin{array}{c} O \\ \uparrow \\ OH - S - O - O - S - OH \\ \downarrow \\ O \end{array}$$

(ii) *These contain – O – O grouping which has been confirmed by the fact that these liberate iodine from iodides and liberate hydrogen peroxide on hydrolysis*

HO H H OH

$$\begin{array}{c} O \qquad\qquad O \\ \uparrow \qquad\qquad \uparrow \\ HO - S - - O - O - - S - OH \\ \downarrow \qquad\qquad \downarrow \\ O \qquad\qquad O \end{array} \rightarrow H2O2 + 2H2SO4$$

HO H

$$\begin{array}{c} O \\ \uparrow \\ HO - S - - O - OH \\ \downarrow \\ O \end{array} \rightarrow H_2SO_4 + H_2O_2$$

(iii) These structure have been confirmed by X-ray diffraction measurements. Definite structural information has been obtained only for the peroxydisulphate ion in its ammonium and caesium compounds. In this ion two SO_4 teterahedra are joined through two oxygen atoms, the S – O and O – O bond distances being 1.50Å and 1.31 Å respectively, and the S – O – O –angel being 128°. The short S – O distance, indicates that the S – O bond has a double bond character.

In the light of the above data, the commonly written structures of or the peroxysulphuric acids appear correct.

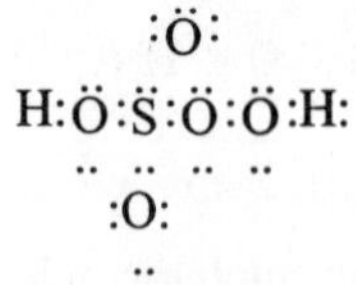

Caro's acid

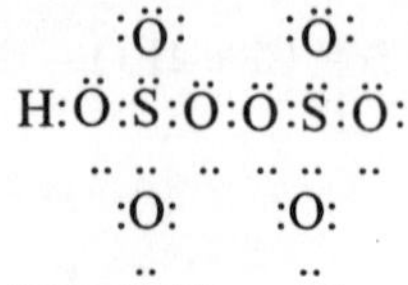

Marshall's acid

Distinction between H_2SO_5, $H_2S_2O_8$ and H_2O_2 :

They may be distinguished from one another by the following tests :

(i) Caro's acid, H_2SO_5, liberates iodine almost instantaneously from potassium iodide.

(ii) Marshall's acid, $H_2S_2O_8$, liberate iodine slowly from potassium-iodine.

(iii) Hydrogen peroxide, H_2O_2, decolourises (reduces) potassium permaganate which is not done be either of the two peroxy sulphuric acids.

6. Sulphur-fluorine compounds containing peroxo groups.

These compound have been made from complex oxofluorides of sulphur. Some examples of these are :

(a) *Bis (Pentafluoro sulphur) peroxide :* It is made by ultraviolet irradiation of the compound F_5SOF. The latter compound is made by treating OSF_2 with an excess of fluorine.

$$\underset{}{OSF_2} + \underset{(\text{Excess})}{2F_2} \rightarrow F_5SOF$$

$$2F_5SOF \xrightarrow{\text{U.V. radiation}} F_5S-O-O-SF_5$$

It is a colourless liquid (b.P. 49°C). It liberates iodine rater slowly from KI.

(b) *Peroxodisulphuryl difluoride :* It is produced in the electrolysis of fluorosulphuric acid. It is also produced in the catalytic fluorination of SO_3 vapours by F_2 when these are passed over AgF2 coated on copper ribbon.

$$2SO_3 + F_2 \xrightarrow{\text{catalyst}} FO_2S-O-O-SO_2F \text{ (B.P.76°C)}$$

It can behave as an oxygenating agent (i), a fluorosulphonating agent (ii) or as both (iii).

(i) $CO + FO_2S - O - O - SO_2F \rightarrow CO_2 + S_2O_5F_2$

(ii) $HgO + 2FO_2S - O - O - SO_2F \rightarrow Hg(SO_3F)_2 + O_2 + S_2O_5F_2$

(iii) $2SOClF + 3FO_2S - O - O - SO_2F \rightarrow 4S_2O_5F_2 + Cl_2$

It liberates iodine rapidly from aqueous KI at room temperature.

7. Per-acids of Chromium

(a) *Chromium Peroxide :* A blue colour is produced when an acidified solution of chromate is treated with hydrogen peroxide. On adding ether, the blue compound remains in etheral solution. After sometimes the blue colour disappears indicating that the compound responsible for the blue colour is unstable.

Wided in 1897 assumed that the blue compound was perchromic acid having the molecular formula $HCrO_5$ and assigned to it the structural formula I given below, in which chromium was represented as being heptavalent.

In all the peracid formations, no change of valency of the central atom is involved during reaction with hydrogen peroxide. In Wided's formula, chromium is represented as being heptavalent while it is hexavalent in chromates.

This and other discrepancies attaching to Wided's formula led Schwarz and Giese in 1932 to reinvestigate the problem. These workers showed that the blue compound is a peroxide CrO_5 and not the acid. They assigned to it the structure II in which the chromium atomic shown as being hexavalent.

$$Cr_2O_7^{2-} + 4O_2^{-2} + 10H^+ \rightarrow 2CrO_5 + 5H_2O$$

Chromium peroxide combines with organic bases like pyridine, piperidine and trimethylamine to form compounds that are insoluble in water but soluble in organic solvents such as benzene. Schwarz and Giese showed that these compounds have the structure (III) in which molecule of the base (pyridine) is co-ordinated with the chromium atom by means of the lone pair of electrons on the nitrogen atom.

(b) *Blue potassium peroxychromate, $K_2Cr_2O_{12}.2H_2O$:* The blue etheral solution of chromium peroxide reacts with alcoholic alkalies, forming blue coloured salts called blue perchromates. This salt decomposes at room temperature to form $K_2Cr_2O_7$.

$$2CrO_5 + 2O_2 + 2KOH \rightarrow K_2Cr_2O_{12} + 2H_2O$$

When the blue crystals come in contact with alkali or dilute acid or get a shock or they are heated, they decompose to regenerate $(CrO_4)^{-2}$.

$$2[Cr_2O_{12}]^{-2} + 4OH^- \rightarrow 4[CrO_4]^{-2} + 2H_2O + 5O_2$$

$$(Cr_2O_{12})^{-2} + 8H^+ \rightarrow 2Cr^{+3} + 4H_2O + 4O_2$$

The peroxy acids corresponding to these salts are to stable in the free state.

The potassium salt has the empirical formula KH_2CrO_7. Schwarz and Giese showed, by simultaneous measurement of the loss of oxygen and water on dehydration, that the salt contains water of crystallisation. Hence the formula should be written as $KCrO_6 . H_2O$.

Further, they proved that it must contain 21/2 peroxy groups for each atom of chromium. So the empirical formula must be doubled. Thus, we get $K_2Cr_2O_{12} . H_2O$ as the molecular formula of the salt. Schwarz and Giese have assigned to this salt the structure (IV)

In this structure, chromium atom is shown as being hexavalent,

The orbital diagram of $[Cr_2O_{12}]^{2-}$ ion may be represented.

(c) *Red Perchromates, K_3CrO_8* : Another series of salts, called red perchromates, have been obtained by the action of hydrogen peroxide on alkaline solution of chromates below 0°C.

$$2CrO_5 + 3H_2O_2 + 6KOH \xrightarrow[(p\varepsilon\delta)]{} 6H_2O + 2K_3CrO_8$$

These possess the empirical formula M_3CrO_8 and have been shown to contain 31/2 peroxo groups for each atom of chromium. Hence as in the case of blue perchromates, the formula must be doubled and we get $M_6Cr_2O_{16}$. The potassium salt has been assigned the following structure :

The orbital diagram of $[Cr_2O_{16}]^{6-}$ may

(d) *Red ammonium peroxychromate $(NH_4)_3\,CrO_5$* : It is prepared by the action of 30% hydrogen peroxide on an ammoniacal solution of ammonium chromate.

The ammonium salt has been shown by conductivity and freezing point measurement to have the monomeric formula and the salts are also isomorphous with pervanadates of the type M_3VO_5. They are, therefore, formulated as derivatives of pentavalent chromium, a conclusion which is supported by the paramagnetic moment. The crystals of this salt are stable at ordinary temperature but they decompose at 170°C and explode at high temperatures.

(e) *Brown perchromate* $CrO_4 3NH_3$: When ammoniacal solutions of chromates are treated with hydrogen peroxide at 0°C, then warmed and cooled, $CrO_4.3NH_3$ separates. Alternatively, the red ammonium perchromate $(NH_4)_3$ CrO_5 may be treated with a 10% ammonia solution at 40°C to yield CrO_4. $3NH_3$.

CrO_4. $3NH_3$ is brown in colour and has the structure :

8. Peracids of Nitrogen

Peracids of nitrogen are :

(a) Pernitrous Acid (HO – O – N= O)

(b) Pernitric Acid (HNO_4)

Both these acids cannot be isolated in the free state.

(a) *Pernitrous Acid* : When sodium nitrite dissolved in 50% hydrogen peroxide is acidified by hydrochloric acid, a light brown colour is produced which disappears with the evolution of oxygen. This brown colour is due to the formation of pernitrous acid.

$$HCl + NaNO_2 \rightarrow HNO_2 + NaCl$$

$$HONO + HOOH \rightarrow HOONO + H_2O$$

When sodium hydroxide is added to the brown solution of pernitrous acid, a bright yellow colour is produced which may persist for 12 hours.

$$NaOH + HOONO \rightarrow NaOONO + H_2O$$

The yellow colour is due to the formation of sodium salt of pernitrous acid. Evidences from chemical reactions show that free radicals are produced due to the decomposition of pernitrous acid.

$$HO - O - N = O \rightarrow HO + NO_2$$

Following examples will favour the formation of free radicals :

(i) The polymerisation of methyl acrylate takes place which is known to be free radical initiated.

(ii) With pernitrous acid, aromatic hydrocarbons form o – hydroxy and m-nitro derivatives which are known to be free radical initiated.

Structure

The structure of pernitrous acid is

$$HO - O - N = O.$$

(b) Pernitric Acid

Preparation

Friederich prepared this acid in pure state by dissolving nitrogen pentoxide in anhydrous hydrogen peroxide.

$$N_2O_5 + H_2O_2 \rightarrow HNO_3 + HNO_4$$

(ii) Rachig reported that pernitric acid is formed in dilute solution by oxidation of nitrous acid with 3% solution of hydrogen peroxide.

$$HNO_2 + 2H_2O_2 \rightarrow HNO_4 + 2H_2O$$

In the above method, free acid could not be isolated because it is extremely unstable. However, a number of it stable salts have been reported.

Properties

(1) It is unstable and cannot be isolated in the free state.

(ii) Its salts are, however, stable but easily decomposed by water.

(iii) Pernitric acid solution liberates bromine from a solution of potassium bromide.

(iv) It is an oxidising agent. It oxidises aniline to nitrobenzene.

Structure

Based on recent X-ray studies, the structure of pernitric acid be represented as follows :

$$\begin{array}{ccc} HO - O - \underset{\substack{\downarrow \\ O}}{N} = O & \text{or} & H:\ddot{\underset{..}{O}}:\ddot{\underset{..}{O}}:\underset{\substack{\ddot{} \\ :\ddot{\underset{..}{O}}:}}{N}::\ddot{O}: \end{array}$$

(c) Potassium Per Oxy Nitrate, KNO_4

It can be prepared by treating potassium nitrate with sodium peroxide in an alcoholic solution.

$$KNO_3 + Na_2O_2 + H_2O \rightarrow 2NaOH + KNO_4.$$

Potassium peroxynitrate is fairly stable and is easily decomposed by water.

9. Peracids of Phosphorous

Two peracids are known :

(i) Peroxy monophosphoric Acid (H_3PO_5)

(ii) Peroxydiphoshoric Acid ($H_4P_2O_8$).

These two acids are analogous to peroxymonosulphuric and peroxydisulphuric acids. These are regarded as derived from orthophosphoric acid (H_3PO_4) and pyrophosphoric acid ($H_2P_2O_7$) by the replacement of one oxygen atom by the peroxy group [– O – O –].

(a) Peroxymonophosphoric Acid, H_3PO_5

Preparation

When phosphorus pentoxide is treated with 30% hydrogen peroxide at low temperature the reaction mixture is diluted and cooled, a solution of peroxymonophosphoric acid is obtained.

$$P_2O_5 + 2H_2O_2 + H_2O \rightarrow 2H_3PO_5$$

Properties

This acid cannot be isolated but its solution produced in the above reaction shown strong oxidising properties. It oxidises manganese (II) salts to permanganate at low temperature.

$$Mn^{+2} \rightarrow MnO_4^-$$

It also liberates iodine immediately from acidified potassium iodide.

Structure

The acid shows structural similarity with H_2SO_5.

The structure of peroxymon

ophosphoric acid can be represented as :

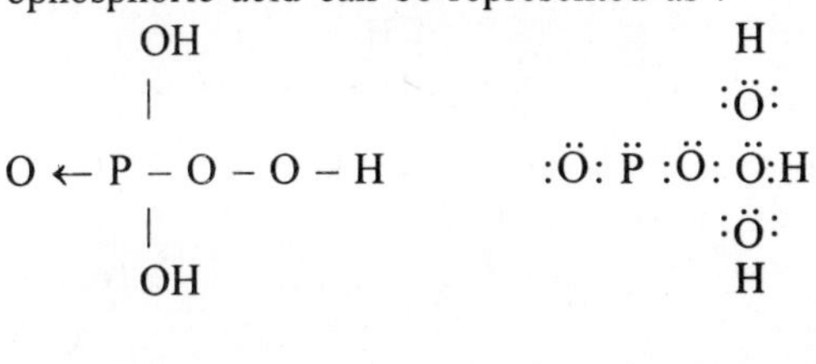

(b) Peroxydiphosphoric Acid, $H_4P_2O_8$

Preparation

(i) It can be prepared by the action of hydrogen peroxide on mate- or pyrophoric acid.

$$2HPO_3 + H_2O_2 \rightarrow H_4P_2O_8$$

(ii) When a concentrated solution of potassium dihydrogen phosphate (KH_2PO_2) is electrolysed in the presence of a high concentration of potassium fluoride, a mixture of relatively stable potassium peroxy diphosphate in addition to the unstable peroxymonophosphoric acid is obtained.

The solution from anodic chamber of evaporation yields potassium peroxydiphosphate ($K_4P_2O_8$).

Properties

(a) A dilute solution of peroxydiphosphoric acid remains stable for a long time.

(b) It shows strong oxidising properties. Examples are :

(i) It liberates iodine from potassium iodide.

(ii) It oxidises aniline to nitrosobenzene and nitrobenzene.

(b) Its solution gets hydrolysed in acidic solution to form peroxymonophosphoric acid and then to hydrogen peroxide.

$$H_4P_2O_8 + H_2O \rightarrow H_3PO_5 + H_3PO_4$$

$$H_3PO_5 + H_2O \rightarrow H_3PO_4 + H_2O_2$$

Structure

The properties of peroxydiphosphoric acid and its salts are similar to peroxydisulphuric acid and its salts. Therefore, the structure of peroxydiphosphoric acid may be written as :

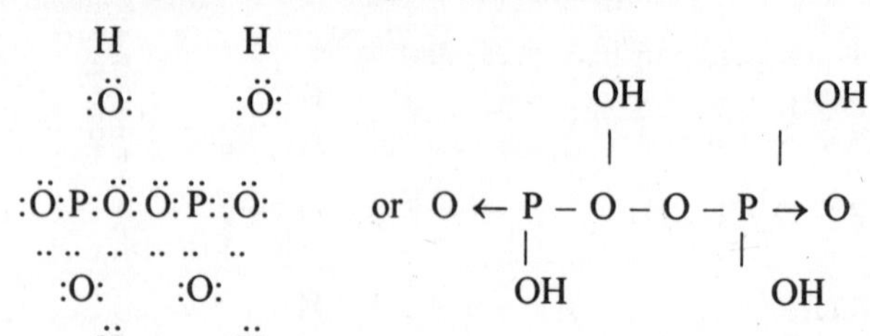

10. Peroxyacids of Vanadium Niobium and Tantalum

$H_4V_2O_{11}$, (K_3NbO_8), K_3TaO_8) are only known as alkali metal salts. When H_2O_2 reacts with vanadyl sulphate a red column due to peroxy vanadyl cation is obtained.

On adding excess of H_2O_2 the colour gets converted to yellow because of the formation of $[VO_2(O - O)_2aq]^{2-}$ ion. This ion is stable in weakly acidic solution and also in weakly alkaline solution.

We can prepare KH_2 $[VO_2(O - O)_2.]$. H_2O by reaction of alkaline solution of vanadium pentoxide and H_2O_2. The salts of the type Na_3M $(O - O)_4$ [M = V, Nb, Ta] are obtained when H_2O_2 reacts with sodium ortho vanadate niobates or tantalates. They lose oxygen even at room temperature.

11. Peracids of Carbon

By analogy with permonosulphuric acid and perdisulphuric acid, the two percarbonic acids may be expected:

OH | O – COOH	O – COOH | O – COOH
Permonocarbonic acid	Perdicarbonic acid

(a) *Peromonocarbonic Acid* : It is not known in the free state. However, its two salts viz. sodium and barium permonocarbonates (Na_2CO_4, and $BaCO_4$) well known.

Preparation

(i) The barium and potassium salts are formed when carbon dioxide is absorbed by sodium peroxide and barium peroxide respectively.

$$Na_2O_2 + CO_2 \rightarrow Na_2CO_4$$

$$BaO_2 + CO_2 \rightarrow BaCO_4$$

(ii) Sodium permonocarbonate may be obtained by the action of phosgene on sodium peroxide.

$$2Na_2O_2 + COCl_2 \rightarrow Na_2CO_4 + 2NaCl + 1/2O_2$$

Properties

The freshly prepared salts of permonocarbonic acid contain water of crystallisation. When dissolved in water, the solution so obtained he haves as if it is a mixture of metal carbonate and hydrogen peroxide. Salts can liberate iodine from a solution of potassium iodide and hydrogen peroxide can be extracted by shaking the aqueous solution with ether.

Structure

The acid and its sodium salt are assigned the following structures :

$$\underset{\text{Sodium permonocarbonate}}{NaO-\overset{}{\underset{\underset{O}{\|}}{C}}-O-O-H} \qquad \underset{\text{Permonocarbonic acid}}{HO-C-\underset{\underset{O}{\|}}{O}-H}$$

(b) *Perdicarbonic Acid :* Perdicarbonic acid is formed in an unstable solution by the action of phosphoric acid on etheral solution of potassium percarbonate.

However, the salts of perdicarbonic acid are fairly stable and are of two types.

Perdicarbonates-Type I : When the electrolysis of cold saturated solution of potassium carbonate or bicarbonate at –12°C using a platinum anode is carried out, a bluish white amorphous ppt of the potassium perdicarbonate is deposited on the anode.

$$2K_2CO_3 \rightarrow 2K^+ + 2KCO_3^-$$

At Cathode,

$$2K^+ + 2e \rightarrow 2K \xrightarrow{+2H_2O} 2KOH + H_2$$

AT Anode.

$$2KCO_3^- \rightarrow 2KCO_3 + 2e$$

$$2KCO_3 \rightarrow \underset{\text{Perdicarbonate}}{K_2C_2O_6}$$

Similarly, the sodium salt has been prepared.

Properties

(i) It is pale blue ppt., which is fairly stable at ordinary temperature.

(ii) On heating, it forms potassium carbonate.

$$2K_2C_2O_6 \rightarrow 2K_2CO_3 + 2CO_2 + 2O_2$$

(iii) Its solution is fairly stable in ice-cold water but is slowly hydrolysed on raising the temperature forming bicarbonate.

$$K_2C_2O_6 + 2H_2O \rightarrow 2KHCO_3 + H_2O_2$$

(iv) If liberates carbon dioxide when reacted with dilute sulphuric acid.

$$K_2C_2O_6 + H_2SO_4 \rightarrow K_2SO_4 + H_2O_2 + 2CO_2$$

(v) It reduces $KMnO_4$ to colourless $MnSO_4$.

$$5K_2C_2O_6 + 8H_2SO_4 + KMnO_4 \rightarrow MnSO_4 + 6K_2SO_4 + 10CO_2 + 8H_2O + 5O_2$$

(vi) It liberates iodine from a cold neutral solution of potassium iodide, oxidises lead sulphide to lead sulphate and bleaches indigo.

$$K_2C_2O_6 + 2KI \rightarrow 2K_2CO_3 + I_2$$

$$4K_2C_2O_6 + PbS \rightarrow 2K_2CO_3 + 4CO_2 + PbSO_4$$

(vii) It converts PbO_2 into $PbCO_3$

$$PbO_2 + K_2C_2O_6 \rightarrow K_2CO_3 + PbCO_3 + O_2 \uparrow$$

Structure

Perdicarbonic acid and potassium perdicarbonate have been assigned the following structures :

These structures are similar to perdisulphuric acid and have been confirmed by X-ray studies.

Perdicarbonates-Type II : These salts are isomeric with perdicarbonates of type-I and can be best represented by sodium perdicarbonate $Na_2C_2O_6$.

Sodium perdicarbonate may be obtained by passing carbondioxide into an ice-cold solution of sodium peroxide in absolute alcohol.

$$Na_2O_2 + 2CO_2 \rightarrow Na_2C_2O_6$$

It differs from the perdicarbonates of type-I in the following points :

(i) It liberates only 50% of the theoretical amount of iodine from the neutral solution of potassium iodide.

(ii) It loses only one-half of its active oxygen as oxygen gas.

On the basis of these two differences from perdicarbonates of type I, the structure of sodium perdicarbonate has been assigned the unsymmetrical formula given below :

Similarly, potassium perdicarbonate may be prepared. Its chemical properties and structures are similar to sodium perdicarbonate.

(c) Perhydrates of Carbonates

Perhydrates of carbonates contain hydrogen peroxide of crystallisation. The most important compound of this type is sodium carbonate perhydrate, Na_2CO_3, H_2P_2 $1/2H_2o$.

It is obtained as precipitate by the action of hydrogen peroxide on sodium carbonate and subsequent addition of absolute alcohol. The sodium carbonate perhydrate is stabilised by the addition of sodium or magnesium silicate and is used in washing powder, etc.

Perhydrates of carbonates liberate iodine very slowly from the neutral solution of potassium iodide. When perhydrates are shaken with ether, their hydrogen peroxide goes into the ether layer.

12. Pertitanic Acid HO – Ti – O – $(OH)_3$

Preparation

(i) It can be obtained by the reaction of H_2O_2 and ammoniacal solution of titanyl sulphate $TiOSO_4$ in aqueous alcohol.

(ii) It can also be obtained by treating H_2O_2 with freshly precipitated titanium hydroxide.

Properties

It is a yellow coloured crystalline solid and loses even at room temperature, both in dry and mist state.

Peroxytitanates : Hydrogen peroxide reacts with alkaline titanyl sulphate to form this compound $K_2TiO_2(SO_4)_2$.

It is an orange coloured compound. It is a true per acid salt because it liberates iodine in buffered (Riesenfeld test) solution. Peroxytitanates $(NH_4)_3$ $[TiO_2F_5]$, $Na_2[TiO_5]$. $3H_2O$ are obtained by the action of hydrogen peroxide on titanium salts. $K_4TiO_4.6H_2O$ is obtained when pertitanic acid reacts with KOH. $K_4TiO_4.6H_2O$ is a colourless solid and does not liberate iodine in buffered Riesenfeld test.

13. Peroxyacids of Zirconium, Hafnium and Thorium

Preparation

$MO_3.2H_2O$ or $M(OH)_3.OOH$.

(i) Hydrogen peroxide reacts with zirconyl or Hafnyl salts to give white gelatenous precipitate of $MO_3.2H_2O$ (M = Zr or Hf).

(ii) On treating thorium nitrate solution with 30% hydrogen peroxide a mixture of $Th(OH)_2(OOH)_2$ and $Th(OH)_3(OOH)$ is obtained. This changes into $Th(OH)_3(OOH)$ after some time.

Properties

(i) These are white solids.

(ii) The stability of these compounds decreases with increase in atomic weight of metal.

(iii) $Zr(OH)_3.OOH$ dissolves in KOH in the presence of H_2O_2 and forms $K_4ZrO_4.4H_2O_2.2H_2O$. This reaction is not given by Hf and Th.

14. Peracids of Boron

The free perboric acid (HBO_3) has not been isolated due to its instability. However, its salts are known.

(I) *Sodium Metaborate Perhydrate $NaBO_2.\ H_2O_2.\ 3H_2O$* : This salt was formerly regarded as true perborate having the formula $NaBO_3\ 4H_2O$ and was erroneously called sodium perborate. But it does not immediately liberate iodine from neutral potassium iodide solution. It means that it cannot be a true persalt. However, it is now accepted that $NaBO_3.\ 4H_2O$ is a metaborate containing hydrogen peroxide of crystallisation.

$$NaBO_2\ .\ H_2O_2.\ 3H_2O$$

Preparation

(i) It is prepared by adding hydrogen peroxide to solution of orthoborate of sodium.

$$Na_3BO_3 + H_2O_2 \rightarrow NaBO_3 + 2NaOH$$

(ii) It may also be prepared by treating borax solution or boric acid with an equal amount of sodium hydroxide and twice hydrogen peroxide.

$$Na_2B_4O_7 + 2NaOH + 4H_2O_2 \rightarrow 4NaBO_3 + 5H_2O$$

$$H_3BO_3 + NaOH + H_2O_2 \rightarrow NaBO_3 + 3H_2O$$

(iii) Technically, it is manufactured by the electrolysis of a borax solution containing sodium carbonate in the ratio 1:3 using platinum gauze electrodes at low temperature. After an hour the solution is allowed to cool. The yield is half the quantity of borax originally taken. It is washed with ice cold water and dried to yield colourless crystalline salt of $NaBO_2$. H_2O_2. $3H_2O$

(iv) It can also be prepared by the action of sodium percarbonate on sodium metaborate

$$NaBO_2 + Na_2CO_4 \rightarrow NaBO_3 + Na_2CO_3$$

(v) It can also be obtained by treating boric acid with NaOOH or Na_2O_2

$$NaOOH + H_3BO_3 \rightarrow NaBO_3 + 2H_2O$$

$$Na_2O_2 + 4H_3BO_3 \rightarrow Na_2B_4O_8 + 6H_2O$$

$$Na_2B_4O_8 + HCl + 4H_2O \rightarrow NaBO_3 + NaCl + 3H_3BO_3$$

Properties

(i) It is a colourless crystalline salt having a somewhat low solubility.

(ii) The dry compound is stable in air. It melts between 150 – 200 °C with the evolution of oxygen.

$$2NaBO_2 \,.\, H_2O_2 \,.\, 3H_2O \rightarrow 2NaBO_2 + O_2 + 5H_2O$$

(iii) Aqueous solution of this behaves like a mixture of borax an hydrogen peroxide.

Structure

This compound has the formula $NaBO_3$. H_2O. As this compound does not liberte iodine from neutral potassium iodide solution, it means that it is not a true per-salt. As in aqueous solution, it behaves like a mixture of borax and hydrogen peroxide; it means that $NaBO_3$. $4H_2O$ is a perhydrate. *i.e.,* a compound containing hydrogen peroxide of crystallisation. On this basis, it is regarded as sodium metaborate perhydrate, $NaBO_2.H_2O_2.3H_2O$.

The $NaBO_2.H_2O_2.3H_2O$, is in fact $Na_2[B_2(O_2)_2(OH)_4]$. $6H_2O$ containing the cyclic ion (Fig. 8). The O – O distance is 1.47Å as in H_2O_2, reveals that speculation about the structure of the compound is less profitable than X-ray crystallographic samination.

Uses

It is used in disinfectant preparation in bleaching and in soap powders.

True Perborate $NaBO_3.4H_2O$

Preparation : An example of this is sodium metaborate having the general composition $NaBO_3$. $4H_2O$. It is prepared by the action of boric acid on sodium hydroperoxide, NaO_2H.

$$NaO_2H + H_3BO_3 \rightarrow NaBO_3 + 2H_2O$$

Structure

As $BaBO_3$ liberates iodine from KI solution, it is true persalt, *i.e.,* a compound containing peroxy linkage – O – O –.

Uses :True sodium perborate is used.

(i) as a mouth-wash, and

(ii) as a bleaching and oxidising agent in cosmetics and soap textiles.

Borate Peroxyhydrate

Per Borax

It is one of best known peroxyhydrates. ($Na_2B_4O_7$. H_2O_2. $9H_2O$). It can be crystallised by dissolving boric acid and sodium peroxide in water.

$$Na_2O_2 + 4H_3BO_3 \rightarrow Na_2B_4O_7.\ H_2O_2 + 5H_2O$$

Perborax changes to borate peroxyhydrates on crystallization from water.

$$Na_2B_4O_7\ .\ H_2O_2 + 4H_2O \rightarrow Na_2BO_2.\ H_2O_2 + NaH_2BO_3 + 2H_3BO_3$$

Borate peroxyhydrates give H_2O_2 in aqueous solution. These are used to prepare washing and bleaching agents for wood, silk, etc. Borate peroxyhydrates having 15% or more of active oxygen are explosive and should be handled carefully.

Care must be taken in handling or working with borate peroxyhydrates containing more than 15% of active oxygen since they may decomposed explosively as result of friction.

Actually speaking there if no appreciable difference between borate peroxyhydrates and peroxyborates particularly from the point of view of their properties.

15. Peracids of Cerium

Two important peracids of cerium are :

(i) Peroxymonocerous acid, and

(ii) Peroxymonoceric acid.

Percerous Acid

It is prepared by the reaction of H_2O_2 and $Ce(OH)_2$ at 5.8 – 9.0

$$Ce(OH)_3 + H_2O_2 \rightarrow OH - O - Ce(OH)_2$$

The acid so formed changes to perceric acid.

$$3HO - O - Ce(OH)_2 + H_2O \rightarrow \underset{\text{Perceric acid}}{2HO - O - Ce(OH)_3} + Ce(OH)_3$$

Perceric acid reacts with cerous hydroxide form ceric hydroxide.

$$HO - O - Ce(OH)_3 + 2Ce(OH)_3 + H_2O \rightarrow 3Ce(OH)_4$$

These peracids have been assigned the following structures.

16. Peroxy Acids of Molybdenum and Tungsten

These peroxy acids are obtained by the action of H_2O_2 with corresponding normal or poly acid.

Properties

Peroxy molybdic acid ($HOOMoO_2$. OH) or (H_2MoO_8) and peroxy tungustic acid ($H_2W_2O_{11}$) are crystalline solids and are yellow coloured.

Peroxy molybdic acid (H_2MoO_8) forms red salts K_2MoO_8 and $ZnMoO_2$. $4NH_3$. It is only in such salts that this acid is known. Its explosive salts like K_2MoO_6, $K_2MoO_8$and $(NH_4)Mo_2O_{11}$. $2H_2O$ can be obtained from solutions.

17. Peroxyuranic Acid

Various peruranates are obtained by the action of alkaline uranyl nitrate solution with H_2O_2. Depending upon the alkalinity different products are obtained.

UO_4. $2H_2O$, a stable peroxide is obtained in neutral solution which disproportonates in alkaline solution.

$$3UO_4 + 4KOH \rightarrow 2UO_3 + K_4UO_8 + 2H_2O$$

$K_2U_2O_{20}$ and K_2UO_8 are formed from strongly alkaline solutions. Uranyl perurante $[UO_2][UO_8]$ xH_2O (x = 6 or 8) are stable, yellow compounds. These are obtained by the reaction of H_2O_2 and a neutral solution of uranyl nitrate.

PROPERTIES OF IODINE PSEUDOHALOGENS

Halogens have a tendency to combine with pseudo halogens to form a new series of compounds known as halogen-halogenoids. A number of such compounds like fluorine azide $(FN)_3$, chlorine azide (ClN_3), bromine azide (BrN_3), iodine azide (IN_3), cyanogen halides (CNCl, CNBr and ICN) and thiocyanogen and oxycyanogen halides are known.

1. *Fluorine azide, FN_3 :* It is a greenish yellow gas, obtained by the reaction of hydrazoic with fluorine in a stream of nitrogen.

   When fluorine azide is cooled, it is liquefied at –82°C and freezed at –152°C to a greenish yellow solid compound. On heating, fluorine azide decomposes forming nitrogen and a new compound of formula N_2F_2.

   $$4FN_3 \rightarrow 4N_2 + 2N_2F_2$$

2. *Chlorine azide ClN_3 :* It is obtained by adding acetic acid to a mixture of sodium azide and hypachlorous acid in aqueous solution.

   $$NaN_3 + H^+ \rightarrow HN_3 + Na^+$$

   $$HN_3 + HOCl \rightarrow ClN_3 + H_2O.$$

   It boils at –15°C and melts at –100°C. It decomposes to nitrogen and chlorine at 400°C and 2mm pressure. With ammonia, it reacts as shown in the following equations :

   $$ClN_3 + 2NH_3 \rightarrow NH_2Cl + NH_4N_3$$

   $$3ClN_3 + 8NH_3 \rightarrow N_2 + 3NH_4Cl + 3\ NH_4N_3$$

3. *Bromine Azide; BrN_3 :* It is prepared by the action of bromine on silver azide or sodium azide.

   $$Br_2 + NaN_3 \rightarrow BrN_3 + NaBr$$

   $$Br_2 + AgN_3 \rightarrow BrN_3 + AgBr$$

   Bromine azide is an orange liquid which freezes to a dark red solid at –45°C. However, its hydrolysis may be represented as :

   $$BrN_3 + H_2O \rightarrow HBrO + HN_3$$

4. *Iodine Azide, IN_3 :* It is a light yellow solid prepared by the action of iodine on mercury (I) azide in ether or benzene.

   $$Hg_2(N_3)_2 + 2I_2 \rightarrow 2HgI + 2IN_3$$

When dissolved in water, it decomposes to iodide, iodate and azide ion.

$$3IN_3 + 2H_2O \rightarrow 2HI + HIO_3 + 3HN_3$$

5. *Cyanogen Halides [CNCl, CNBr, CNI]:* All these halides can be prepared by action of desired halide on alkali cyanide.

$$X_2 + HCN \rightarrow CNX + MX$$

where $H = Cl, Br \text{ or } I$

All the cyanogen halides except CNF undergo trimerisation to form cyanuric trihalides. However, this tendency to polymerise may be related to the purity of the compound.

It freezes at –6.9°C and reacts with water in the presence of hydrochloric acid.

$$CNCl + 2H_2O \rightarrow CO_2 + NH_4Cl$$

Hydrochloric acid is added to increase the speed of reaction. With sodium azide, it yields, the inter-pseudo halogen.

$$CNCl + NaN_3 \rightarrow CNN_3 + NaCl$$

When CNCl and CNBr are boiled with KOH, potassium bromide and chloride are obtained.

$$CNBr + 2KOH \rightarrow KOCN + KBr + + H_2O$$

$$CNCl + 2KOH \rightarrow KOCN + KCl + H_2O$$

With ICN, it gives iodate and cyanide

$$ICN + 2KOH \rightarrow KCN + KIO + H_2O$$

$$3KIO \rightarrow KIO_3 + 2KI$$

If these cyanogens are kept for some time, they undergo polymerisation

$$3CNCl \rightarrow [CNCl]_3$$

$$3CNBr \rightarrow [CNBr]_3$$

$$3CNI \rightarrow [CCI]_3$$

These polymerised cyanogen halides have cyclic structure.

THIOCYANOGEN HALIDES AND OXYCYANOGEN HALIDES

These are prepared by treating led thiocyanate with appropriate halogen

$$Pb(SCN_2) + 2X_2 \rightarrow (XSCN)_2 + PbX_2$$

where X is Cl, Br or I.

Oxycyanogen halide may be prepared by treating halogen with silver cyanate.

$$AgOCN + X_2 \rightarrow AgX + XOCN$$

where X = Cl, Br or I

All the oxycyanogen halides exist as dimers and their structures may be represented by a common formula as depicted below :

$$X_2N - \overset{\overset{\displaystyle O}{\|}}{C} - N = C = O$$

INTERHALOGENOIDS

Such compounds are formed by the combination of two pseudo-halogens. These compounds are analogous to the interhalogen compounds.

Cyanogen derivatives of other pseudo-halogens have been reported. Examples are cyanogen thiocyanate, (CNSCN), cyanogen azide (CNH_3) and cyanogen azido-di-thiocarbonate ($CNSCSN_3$).

We shall now discuss some of the interhalogenoids :

(a) *Cyanogen thiocyanate* : It is best prepared by the action of mercuric cyanide on thiocyanogen.

$$Hg(CN)_2 + 2(SCN)_2 \rightarrow Hg(SCN)_2 + 2CNSCN$$

Another method involves the action of alkali thiocyanate on cyanogen bromide

$$KSCN + CN.Br \rightarrow KBr + CN.SCN$$

Cyanogen thiocyanate is a crystalline volatile solid having melting point as 61°C.

(b) *Cyanogen selenocyanogen* : The existence of this compound was first of all reported by H.P. Kaufmann and F. Kogler. However, it can be obtained by the decomposition of selenocyanogen.

$$(SeCN)_2 \rightarrow Se + CNSeCN$$

It is a crystalline volatile solid having melting point as 36°C.

(c) *Cyanoazide :* It may be best prepared by the action of cyanogen bromide on sodium azide.

$$CNBr + NaN_3 \rightarrow NaBr + CNN_3$$

It is a volatile, crystalline solid melting at 36°C. It is isometric in nature.

$$2CNN_3 \rightleftarrows N \equiv C - N = C(N_3)_2$$

(d) *Cyanogen azido dithiocarbonate :* It may be prepared by treating mercury (II) cyanide with $(SCSN_3)_2$.

$$Hg(CN)_2 + (SCSN_3)_2 \rightarrow Hg(SCN_3)_2 + 3CNSCSN_3$$

It is a white crystalline solid having more solubility in organic solvents than water at 0°C.

However, it undergoes explosion on heating. When kept for some time, it undergoes gradual decomposition according to the following equation :

$$CNSCSN_3 \rightarrow N_2 + S + CN.SCN$$

$$_x(CNSCN) \rightarrow (CN.SCN)_x$$

$$CNSCSN_3 \rightarrow N_2 + 2SCN$$

$$_x(SCN) \rightarrow (SCN)_x$$

With ammonia, the following reaction takes place

$$CN.SCSN_3 \rightarrow 2NH_2 + NH_4SCSN_3 + H_2NCN$$

Cyanogen azide dithiocarbonate is considered to have a structure similar to that of an interhalogen.

$$N \equiv C - S - \underset{\underset{S}{\|}}{C} - N = N \equiv N$$

POLYHALIDES

When halide ions associate with molecules of halogens or interhalogens, univalent ions are obtained. These are known as polyhalide ions and the compounds of these are known as polyhalides.

As in the case of interhalogens, the heavier halogen atom in poylhalides plays the role of central atom. Some examples of polyhalides are :

(i) It has been suggested that the solubility of iodine in water is increased by the addition of iodide ions which is due to the formation of the tri-iodide ion I_3^- (polyhalide ion).

$$I_2 + I^- \rightleftarrows I_3^-$$

The formation of more complex polyhalide ions like I_5^-, I_7^-, I_9^-, have also been confirmed by spectral studies.

(ii) When bromine is added to a metal halide, mixed polyhalide is obtained. This is carried out in the absence of a solvent.

Solvated Polyhalides

Many crystalline polyhalides contain molecules of the solvent or other foreign materials which are essential for the stability of the polyhalides because the whole substance decomposes into a monohalide and a halogen or interhalogen when these foreign molecules are removed.

Such polyhalides are called solvated polyhalides. Examples are : K_3, H_2O, KI_3, $2H_2O$, KI_7, H_2O, CsI_{10}. $2C_6H_6$, $Be(IC_4)_2$. $8H_2O$.

Conditions for the Formation of Polyhalides

(i) The formation of polyhalide ions may occur in the solution or in the solid state.

(ii) Generally, polyhalide formation is restricted to large cations like alkali metal ion, an alkaline earth metal ion, a coordination complex ion such as $[Co(NH_3)_6]^{+3}$, an organic base, etc.

(iii) Polyhalide formation is more common in iodides than the other halides.

(iv) The tendency of polyhalide formation increases with decreasing lattice energy in the original halide.

That is why chlorides form less polyhalides, bromides form more whereas iodides commonly enter into polyhalide formation.

(v) No concrete evidence of isomerism among the polyhalides has been obtained. $CsIBr_2$ exists in two forms which have different colours.

This is not due to different isomers but due to the difference in crystal size.

Classification of Polyhalides

Due to the work of Weers and other, the following classification is possible. Polyhalides are of three types depending upon the type of the polyhalide ions.

(i) *Type I :* It includes compounds giving polyhalide ions of the type X_n^-. Examples are RbI_3, CsI_3, NaI_3, $2H_2O$ KI_3, $2H_2O$. The compound of potassium (KI_3) is of particular interest. Many attempts have been made to obtain the solid KI_3 but one cannot get pure KI_3 but solvated potassium tri-iodide, *i.e.*, KI_3, $2H_2O$.

(ii) *Type II :* It includes most compounds which yield ions of the type $(XX'_n]^-$. These are very common and the most stable is in the non-solvated conditions. It includes salts of the series

$$IBr_2^-, ICl_2^-, ClBr_2^- \text{ and } BrCl_2^-.$$

The only example of material containing $(XX'_6]^-$ type of group is KIF_6. This compound on heating gives iodine pentafluoride at its melting point. Water hydrolyses this compound to yield fluoride and

(iii) *Type III :* Type III includes compounds giving ions of the type $(XX'X"]^-$. These are not very common. Examples are :

$CsFIBe_3$, $RbFICl_3$, CsFICl

KClIBr, RbClIBr, CsClIBr

This is more important because the only known polyhalides containing fluorine belong to it.

In addition to above polyhalide cations like BrF_2^+, ICl_2^+, IF_4^+ are also known which arise out by the unusual dissociation of the interhalogens AB_3 and AB_5.

$$ICl_3 \Leftrightarrow ICl_2^+ + Cl^- \; ; BrF_3 \Leftrightarrow brF_2^+ + BrF_4^- \; ;$$

$$2IF_5 \Leftrightarrow IF_4^+ + IF_6^-$$

Alternative Classification of Polyhalides

The best way of classifying polyhalide ions is that they are classified on the basis of the total number of halogen atoms present in a particular polyhalide ion. Thus these ions are classified as :

(i) Trihalide ions : I_3^- . ICl_2^-, $IBrF^-$, BrF_2^+ , ICl_2^+ , ClF_2^+ etc.

(ii) Pentahalide ions : etc. ICl_4^-, BrF_4^- , I_5^- ,IF_4^+

(iii) Heptahalide ions : I_7^- ,IF_6^- etc.

(iv) Other odd polyhalide ions : I_9^- ,I_{11}^-, Br_9^- etc.

(v) Other even polyhalide ions : I_4^- ,I_8^- etc.

Table 2 : Classification of polyhalide ions.

Trihalide ions	*Pentahalide ions*	*Heptahalide ions*	*Other odd polyhalide ions*	*Other even polyhalide ions*
I_3^-	IC_4^-	IF_6^-	I_9^- , Br_9^-	I_4^-
ICl_2^-	BrF_4^-	I_7^-	I_{11}^-	I_8^-
$IBrF^-$	I_5^-			
BrF_2^+	IF_4^-			
ICl_2^+				
ClF_2^+				

General Methods of Preparation of Polyhalides

(i) It is possible to prepare simple polyhalides like $K^+Cl_3^-$, $NH_4^+I_5^-$ etc. by the action of halogen on metallic halides. For example.

$$Cl_2 + KCl \rightarrow K^+Cl_3^-$$

$$2I_2 + NH_4I \rightarrow NH_4^+I_5^-$$

In some cases the reaction is preformed in a solvent. The solvent used must not react with the reactants, must dissolve the halide and must not solvolyse. For example :

$$\left[(CH_3)_4N^+\right]I^- + \left(\frac{n-1}{2}\right)I_2 \xrightarrow{CH_3COOH} \left[(CH_3)_4N^+\right]I_n^-$$

The value of n has been found to depend on the amount of I_2 used in the reaction.

(ii) Mixed polyhalides like $Cs^+[ClF_4]^-$, $K^+[ICl_2]^-$ etc. are also prepared by the action of an appropriate gaseous halogen or interhalogen on a solid halide in the absence of a solvent . Examples are as follows :

$$Cl_2 + KI \rightarrow K^+[ICl_2]^-; 2F_2 + CsCl \rightarrow Cs^+[ClF_4]^-$$

$$ICl + KCl \rightarrow K^+[ICl_2]^-; ICl + KBr \rightarrow K^+[BrICl]^-$$

$$ICl_3 + KCl \rightarrow K^+[ICl_4]^-; BrF_3 + KF \rightarrow K^+[BrF_4]^-$$

Here too, in some cases a suitable solvent is required for the reaction. For example.

$$Br_2 + CsI(g) \xrightarrow{C_2H_5OH} Cs^+[IBr_2]^-$$

$$ICl + [(CHl3)_4 N]Br \xrightarrow{CH_3COOH} [(CH_3)_4 N]^+ [BrICl]^-$$

(iii) Polyhalides of type such as $M^+[IC_2]$ (M^+ = Li^+, Na^+ etc. and $M^{2+}[ICl_4]_2^-$ (M^+ = Mg^{2+}, Ca^{2+} etc.) can be obtained by adding I_2 to the solution of metallic chloride in HCl followed by saturation with Cl_2. For Example :

$$2M^+Cl^- \text{ (solution in HCl)} + I_2 + 3Cl_2 \rightarrow 2M+[ICl_4]^-$$

$$M^{2+}Cl_2^- \text{ (solution in HCl)} + I_2 + 2Cl_2 \rightarrow M^{2+}[ICl]_2^-$$

$$2Rb^+Cl^- \text{ (solution in HCl)} + I_2 + Cl_2 \rightarrow 2Rb^+[ICl_2]^-$$

(iv) Polyhalides of one type can be converted into the other type by the action of halogens or interhalogens on them, for example :

$$K[ICl_2] + Cl_2 \rightarrow K[ICl_4]; CsBr_3 + I_2 \rightarrow Cs[IBr_2] + IBr$$

$$K[IBr_2] + Cl_2 \rightarrow K[ICl_2]; Br_2; K[ClIBr] + ICl \rightarrow K[ICl_2] + IBr$$

(v) In a few polyhalide acids such as H^+ $[ICl_4]^-$, $H^+[ICl_2]^-$, $H^+[Br_2]^-$ and $H^+[BrCl]^-$ can be prepared by adding a halogen or an interhalogen to an acid. H^+ $[ICl_4]^-$, is obtained as a tetrahydrate, H^+ $[ICl_4]^-$. $4H_2O$ which is dehydrated only on decomposition. The polyhalide acid much as H^+ $[ICl_2]^-$, H^+ $[IBr_2]^-$ and H^+ $[IBrCl]^-$, exist only in solution.

General Physical Properties

(i) *Colour* : All the polyhalides are coloured compounds. The intensity of the colour increases with the increase of the atomic number of the halogen atoms present. Thus the bromoiodides are red and chlorobromides are yellow. Their absorption spectra are characterised by bands in the regions 2000 to 2900 Å and 3400 to 3900 Å.

(ii) They have in general low melting points.

(iii) Polyhalides dissolve in liquids of high dielectric constant, *e.g.*, water, acetones, alcohol, etc. They are generally insoluble in liquids of low dielectric constant except the free halogens and interhalogens. Dissociation in water is generally followed by hydrolysis.

$$I_3^- \rightarrow I^- + I_2.$$

General Chemical Properties

(i) *Stability* : Polyhalides are typically ionic compounds though they tend to decompose on heating .

$$CsI_3 \xrightarrow{\text{Heat}} CsI + I_2$$

$$RbICl_2 \xrightarrow{\text{Heat}} RbCl + ICl.$$

Dissociation is readily affected by raising the temperature but is often appreciable at room temperature or below. Polyhalides containing one iodine atom are normally less dissociated than those containing no iodine or more than one iodine atom.

For a given type of polyhalide, thermal stability increases in the series $Na^+ < K^+ < NH_4^- < Rb^+ < Cs^+$.

It is to be noted, that the size of Na^+, Rb^+ and Cs^+ cation increases from Na^+ to Cs^+. The stability of polyhalide ion increases in the following order as :

$$[BrI]^- < [FIBr]^- < [ClIBr]^- < [I_3]^- < [BrIBr] < [ClICl]^-$$

Sharpness of the melting point is a criterion of thermal stability.

No F_3^- compounds are known because fluorine can not expand its octet due to the absence of d-orbitals in it.

(ii) The polyhalides having ICl_4^- anion dissociate in aqueous solution to metal chloride and ICl_3 : the latter then hydrolyses to iodate, IO_3^-.

$$
\begin{array}{l}
M^{+}[ICl_4]^{-} \rightarrow MCl + ICl_3] \times 5 \\
5ICl_3 + 9H_2O \rightarrow 3HIO_3 + 15HCl + I_2 \\
\hline
5M^{+}[ICl_4]^{-} 9H_2O \rightarrow 3HIO_3 + 15HCl + I_2 \\
5ICl_4^{-} 9H_2O \rightarrow 20Cl^{-} + 3IO_3^{-} + 18H^{+} + I_2
\end{array}
$$

(iii) *Formation of complexes:* Some of the polyhalide ions are able to form complexes with organic donor molecules. Some examples of such complexes are : LiI_3. $4C_6H_6CN$, MI_3, $2C_6H_5CN$ (M = Na or K).

(iv) *Dissolution in liquid halogen or interhalogen :* Polyhalides may dissolve in liquid halogen or interhalogen to give solution from which the original polyhalide may be crystallised on evaporation. The polyuhalides like $K^{+}[IBr_2]$, $NH^{+}{}_4[IBr_2]^{-}$ and $Rb^{+}[IBr_2]^{-}$ form solution in bromine.

(v) *Replacement properties :* A halogen in the polyhalide may be replaced by a more electronegative atom but the central halogen atom is not replaceable. That is why chlorine can not replace central iodine atom in IBr_2^{-} but replaces bromine in IBr_2^{-} to yield ICl_2^{-}.

$$IBr_2^{-} + Cl_2 \rightarrow ICl_2^{-} + Br_2$$

Similarly, bromine may replace iodine. For example I_3^{-} is converted into IBr_2^{-} but not to IBr_3^{-}.

$$I_3^{-} + Br_2 \rightarrow IBr_2^{-} + I_2$$

(vi) *Action of halogens :* A halogen may add on directly to the polyhalides where as this behaviour is not exhibited by interhalogens. For example. ICl_2^{-} reacts with chlorine to give ICl_4^{-}.

$$ICl_2^{-} + Cl_2 \rightarrow ICl_4^{-}$$

(vii) *Oxidants :* ClF_4^{-} and BrF_6^{-} are powerful oxidants which explode with organic solvents. The structures of simple polyhalide ions are known.

Structure and Shape of Trihalide Anions

(Examples : ICl_2^{-}, I_3^{-}, IBr_2^{-}, $IBrF^{-}$ etc.) The tri-halide anions like ICl_2^{-}, I_3^{-}, $IBrF^{-}$ have symmetrical linear shape while the trihalide anions like IBrF$^-$ have unsymmetrical linear shape. The structures of ICl_2^- and I_3^- ions have been considered as follows :

(i) Structure and shape of ICl_2^- ion. This ion is having symmetrical linear shape which results from sp^3d hybridisation of I-atom (central atom). Because of the presence of one unit of negative charge on ICl_2^- ions, I-atom (central atom) can be considered as having eight electrons (instead of sever) in its valence-shell.

The Lewis structure of ICl_2^- ion, $[Cl - I - Cl]^-$ reveals that I-atom is surrounded by two σ-bps and the remaining six electrons remain as three lps on I-atom. Thus, since σ-bps + lps = 2 + 3 = 5, I-atom is sp^3d hybridised in ICl_2^- ion as shown below:

$5s^2$, $5px^2 5py^2 5pz^1$. $5d^o$ (I-atom in its ground state) $\xrightarrow{+e}$

$5s^2$, $5px^2 5py^2 5pz^2$. $5d^o$ (I-atom after having gained one electron)

$\rightarrow$ $5s^2$, $5px^2 5py^2 5pz^1$. $5d^1$ (I-atom in its excited state)

$\xrightarrow{sp^3d \text{ hybridisation}}$ Five sp^3d hybrid orbitals.

All the three basal hybrid orbitals are having lone pairs of electrons.

Now the two axial hybrid orbitals, each of which is having one electron, overlaps with the 3px orbital (singly-filled) of two Cl^- atoms to form two I-Cl σ-bonds. Due to the presence of lps. ICl_2^- ion assumes linear symmetrical shape.

The three lone pairs of electrons (lps) occupy the basal positions of the trigonal bipyramid. Theoretically three lps are able to occupy any three positions of the trigonal bipyramid and thus give rise to three different structures namely (a), (b) and (c) to ICl_2^- ion as depicted.

The most stable structure is that in which the angle between two lps is the largest angle. Now because in structure (a) the angle between two lps is 120° which is the largest, angle, this structure is the most stable, *i.e.,* ICl_2^- ion is having linear shape.

(ii) Structure and shape of I_3^- ion is also having linear shape

In certain cases I–I–I bond angle in I_3^- ion is not equal to 180° and the two I–I bond lengths are unequal (unsymmetrical structure) as depicted below in some ionic salts.

Structure and Shape of Trihalide Cations

(*Examples* : ICl_2^+, BrF_2^+, ClF_2^+). The trihalide cations like ICl_2^+, BrF_2^+, ClF_2^+ are having angular shape which results from sp^3 hybridisation of the central atom.

As an example let us consider the structure of ICl_2^+ ion. In this ion I-atom is the central atom. Because the presence of one unit of positive charge on ICl_2^- ion, I-atom (central atom) can be considered as having six electrons (instead of seven) in its valence shell.

The Lewis structure of ICl_2^- ion, $[Cl - I - Cl]^+$ reveals that I-atom has σ–bps = 2 and lps = 2, and since σ – bps + lps = 4.I⁻. atom is supposed to be sp^3 hybridised as shown below :

$5s^2$, $5px^2 5py^2 5pz^1$. $5d^o$ (I-atom in its ground state) $\xrightarrow{-e}$

$5s^2$, $5px^2 5py^2 5pz^o$. $5d^o$ (I-atom after having lost one electron)

→ $5s^2$, $5px^2 5py^1 5pz^1$. $5d^1$ (I-atom in its excited state)

$\xrightarrow{sp^3d\ \text{hybridisation}} (sp^3)^2\ (sp^3)^2\ (sp^3)^1 (sp^3)^1$

(Four sp^3d hybrid orbitals two of which contain lps and the remaining two are singly-filled).

From sp^3 hybridisation scheme shown above it follows that two sp^3 hybrid orbitals are having lps while each of the remaining two sp^3 hybrid orbitals has one electron. These two singly-filled sp^3 hybrid orbitals make a head-to-head (linear) overlap with singly filled $3p_x$ orbitals on each of the two Cl-atoms ($Cl = 3s^2$. $3px^1 3py^2 3pz^2$) and form two $sp^3(I) - 3px(Cl)$ σ-bonds.

Although, the spatial arrangement of four electron pairs round I-atom has been tetrahedral, due to the presence of two lone pairs of electrons, the shape of ICl_2^+ ion is distorted and becomes angular or V-shape as shown in Fig. 4.7.

The I–Cl bond length and ClICl bond angle found in ICl_2^+ ion have been found to depend on the nature of the anion attached to this cation. For example the values of bond angle and bond length in $[ICl_2]^+[SbCl_2]^-$ have been 92.5° and 2.31 Å respectively whereas these values in $[ICl_2]^+[AlCl_2]^-$ have been 96.7o and 2.28 Å respectively.

STRUCTURE AND SHAPE OF PENTAHALIDE ANIONS

(Examples : I^-_5, ICl^-_4, BrF^-_4). The pentahalide anicns such as I^-_5, ICl^-_4, BrF^-_4 etc. are having square planar shape which results from sp^3d^2 hybridisation of the central atom.

As an example let us consider the structure of ICl_4^- ion. In this ion I-atom is the central atom. Because of the presence of one unit of negative charge on ICl_4^- ion. I-atom (central atom) can be considered as having eight electrons (instead of seven) in its valence shell. The Lewis structure of this ion,

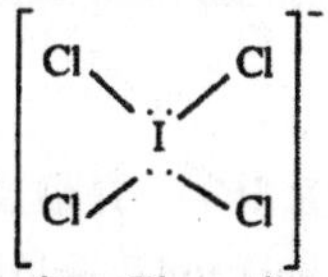

reveals that I-atom is surrounded by four σ-bps and two lps. Thus sine σ –bps + lps = 4 + 2 = 6, I-atom is regarded to be sp^3d^2 hybridised as shown below :

$5s^2$, $5px^2 5py^2 5pz^1$. $5d^o 5d^o$ (I-atom in its ground state) $\xrightarrow{+e}$

$5s^2$, $5px^2 5py^2 5pz^o . 5d^o 5d^o$(I-atom after having gained one electron)→

$5s^2$, $5px^2 5py^1 5pz^1$. $5d^1 5d^1$ (I-atom in its second excited state)

$\xrightarrow{sp^3d \text{ hybridisation}}$ six sp^3d^2 hybrid orbitals.

The two lps occupy the axial hybrid orbitals of the octahedron.

sp^3d^2 hybridisation shown above makes it evident that out of the six hybrid orbitals two axial hybrid orbitals are having tow ps, because so as to have minimum lone pair repulsion, the lone pairs must be as far from each other as possible. Other four hybrid orbitals (equatorial hybrid orbitals) which are singly-filled overlap with the singly-filled 2px orbitals on for Ci-atoms will tend to form four I-Cl σ-bonds.

Although the spatial arrangement of six electron pairs round I-atom has bee octahedral, but because of the presence of two lone pairs of electrons in the axial hybrid orbitals, the shape of ICl_4^- ion gets distorted an becomes square planar.

Structure and Shape of Pentahalide Cations

(Example : IF_4^+*).* In $[IF_4^+]^+[SbF_6]^-$, IF_4^+ cation is having distorted tetrahedral or seasaw geometry with one lone pair of electrons occupying one of the equatorial (*i.e.,* basal) positions of the trigonal bipyramid.

Because of the presence of one unit of positive charge on IF_4^+ ion. I-atom (central atom) may be considered as having six electrons (instead of seven) in its valence shell. Lewis structure of IF_4^+ ion reveals that the central atom (*i.e.*, I-atom) is surrounded by four σ-bps and one lp, *i.e.*, σ-bps + lps = 5.

Hence, since σ-bps + lps = 5 I-atom in IF_4^+ ion is sp^3d hydribidised. Because of the presence of one lone pair of electrons which occupies one of the basal hybrid orbitals, the shape of IF_4^+ ion is distorted and becomes distorted tetrahedral or sea-saw.

STRUCTURE AND SHAPE OF HIGHER POLYHALIDE ION

(Example : I_7^-, I_9^-, I_8^- *etc).* The structure of hepta iodine ion, I_7^- in $[(C_2H_5)_4N]^+I_7^-$ is not having discrete I_7^- ions but rather having three-dimensional array of a linear I_3^- ion and two I_2 molecules in the ratio 1 : 2 ($I_3^- - I_2$). The structure of I_9^- ion in $[(C_2H_5)_4N]^+$ I_9^- is having one I_5^- unit linked with two I_2 molecules. The structure of I_8^{2-} ion in Cs_2I_8 consists of two I_3^- units linked with one I_2 molecules.

IODINE PROPERTIES

In the periodic table, metallic or basic properties of elements increases on descending a group. Thus, in groups IV, V and VI the first elements. C, N and O are non-metals whereas the heavier members Pb, Bi and Po are metals, Similarly one can expect the same trend in group VII, *i.e.*,

(i) Fluorine is the most electronegative element and possessing no basic or metallic properties;

(ii) chlorine shows a slight basic character :

(iii) Bromine shows a comparatively higher basic properties in its complexes;

(iv) Iodine show strong basic character, *i.e.*, the tendency of iodine atoms to form cations.

The basic character of iodine is well established. It must be emphasized that iodine is not a metal yet it is exhibiting basic properties. There are a large number of compounds which contain basic iodine.

Reasons For Basic Properties of Iodine

1. The electronegative character in halogens decreases from fluorine to iodine. Iodine is least electrogegative and exhibits electropositive character.

	F	Cl	Br	I
Electronegativity	4.0	3.0	2.8	2.5
Ionisation Potential in eV.	17.4	13.0	11.8	10.4

2. From the values of ionisation potential, it is evident that in the case of iodine it is easier to lose one electron to form positive iodine.

3. Size of atom increases from fluorine to iodine which further confirms positive character of iodine.

	F	Cl	Br	I
Atomic radius in Å	0.62	0.99	1.14	1.33

From the above values, it can be assumed that iodine can lose more than one electron, *i.e.*, it can form cation with three positive charges, I^{3+}.

4. Ionic radii values increase from fluorine to iodine. Due to this reason, the nucleus of iodine can hold the electrons in the outer most orbit less firmly.

   Therefore, iodine can form cations with one positive charge I^+, and with three positive charges, I^{3+}. Iodine forms two types of salts in which iodine is univalent and trivalent. Iodine is present as cationic constituent in salt-like compounds formed by it.

We shall now discuss evidences for the existence of positive iodine.

Evidences For I^+

(i) When the compounds like ICl, IBr, ICN, INO_3, I_2SO_4, $[Ipy_2]^+ X^-$ ($X^- = NO_3^-$, ClO_4^-), $I(ClO_3)_3 \cdot 2H_2O$, $I(CH_3COO)_3$, ICl_3, IPO_4, $I(NO_3)_3$, $(IO)_2 SO_4$, $H[ICl_3] \cdot 4H_2O$ in their molten state or as their aqueous solution are electrolysed, iodine is liberated at cathode.

The liberation of iodine at cathode proves the existence of cationic iodine in these compounds.

$$2ICl \rightleftharpoons I^+ + ICl_2^-$$

$$ICN \rightleftarrows I^+ + CN^-$$

(ii) ICl acts as a strong electrophilic iodinating agent, converting acetanilide into 4-iodiacetanilide and salicylic acid into 3, 5 di-iodiosalicylic acid.

It again confirms that iodine can form positive ions.

(iii) The iodine is dissolved in an inert solvent. If it is allowed to pass down a cationic ion exchange column, some iodine is retained on the resin.

$$H^+ \text{ Resin} + I_2 \rightarrow I^+ \text{ Resin} + HI$$

The positive ions retained by resin may be eluted with KI solution.

$$I^+ \text{ Resin} + KI \rightarrow K^+ \text{ Resin} + I_2$$

Again, this study confirms that iodine forms positive ion.

(iv) Positive iodine is present in the stable complexes of pyridine :

I (Pyridine) NO_3, I (pyridine)$_2$ NO_3

I (pyridine)$_2$ ClO_4, I (pyridine) acetate and

I (pyridine) benzoate.

When electrolysis of I (pyridine)$_2$ NO_3 in chloroform solution is carried out, iodine is evolved at the cathode.

(v) Spectroscopic obervations are direct and better indicative for I^+. According to F. Aubke (1965), the products obtained by the reaction of iodine with $S_2O_6F_2$ gives colour and characteristics spectrum of I^+.

(vi) When I_2 is dissolved in H_2O, a variety of reactions take place. Two or such reactions are as follows :

(a) $I_2 + H_2O \rightleftharpoons [I(H_2O)]^+ + I^-$

(b) $[I(H_2O)]^+ + H_2O \rightleftharpoons [I(OH)]^0 + H_3O^+$

The formation of $[I(H_2O)]^+$ and $[I(OH)]^0$ in these reactions also confirms the existence of I^+ ion.

COMPOUNDS CONTAINING MONOVALENT IODINE CATION I^+

Compounds having monovalent iodine cation or unipositive iodine I^+ are as follows :

1. *ICl and IBr interhalogen compounds:* For the preparation, properties, uses and structure of these compounds see "Interhalogen Compounds".

2. *Iodine nitrate INO_3 and iodine sulphate I_2SO_4 :* Solutions of these compounds are prepared by passing a solution of HNO_3 and H_2SO_4 respectively in absolute through ion exchanger containing sorbed I^+ ion.

$$I^+R^- + HNO_3 \rightarrow INO_3 + HR$$

$$2I^+R^- + H_2SO_4 \rightarrow I_2SO_4 + 2HR$$

If the aqueous solutions of these compounds are electrolysed, iodine gets liberated at cathode, revealing that both of these salts contain unipositive iodine, I^+.

3. *Complex compounds of $[Ipy_2]^+ X^-$ type $X^- = NO_3^-, ClO_4^-$:* In these compounds I^+ ion is stabilished by coordination. These compounds are prepared by treating Ag (I) salt with I_2 in chloroform in presence of excess of pyridine (py). For example :

$$AgNO_3 + I_2 + 2py \rightarrow AgI(ppt) + [Ipy_2]NO_3$$

It is a colourless crystalline salt having a somewhat low solubility.

The dry compound is stable in air. It melts between 150 – 200 °C with the evolution of oxygen.

$$2NaBO_2 \,.\, H_2O_2 \,.\, 3H_2O \rightarrow 2NaBO_2 + O_2 + 5H_2O$$

Aqueous solution of this behaves like a mixture of borax an hydrogen peroxide.

Complexes of this type exhibit the following properties. We shall consider $[Ipy_2]NO_3$ as an example.

(a) When $[Ipy_2]NO_3$ reacts with acidified KI solution, I_2 gets liberated.

$$I^+ \text{ (From complex compound)} + I^- \text{ (from KI)} \rightarrow I_2$$

I_2 liberated in the above reaction can be quantitatively estimated with sod. thiosulphate, $Na_2S_2O_3$

$$2Na_2S_2O_3 + I_2 \rightarrow Na_2S_4O_6 + 2NaI$$

(b) Electrolysis of $[I_{py2}]\,NO_3$ in chloroform liberates iodine at cathode.

$$2I^+ \text{ (From the complex compound)} + 2e \rightarrow I_2 \text{ (At cathode)}$$

Reactions (a) and (b) reveal the presence of I^+ ion in the complex compound, $[I_{py2}]\ NO_3$.

(c) $[I_{py2}]\ NO_3$ is hydrolysed by water as follows:

$$5[I_{py2}]\ NO_3 + 3H_2O \rightarrow 5HNO_3 + 10py + HIO_3 + 2I_2$$

(d) Treatment of $[I_{py2}]\ NO_3$ with NaOH yields the base $[I_{py2}](HO)$ which is converted into its anhydride.

$$[I_{py2}]\ NO_3 + NaOH \rightarrow [I_{py2}]^+(OH)^- \text{ (base)} + NaNO_3$$

(e) Reaction of $[I_{py2}]\ NO_3$ with phenol, C_6H_5OH yields triodophenol, $C_6H_2I_3\ OH$

$$C_6H_5OH + 3[I_{py2}]\ NO_3 \rightarrow C_6H_2I_3\ OH + 3HNO_3 + 6py$$

(f) Noble metals like Au are dissolved in chloroform solution of the complex compound.

$$3[I_{py2}]\ NO_3 + Au \rightarrow Au^{3+} + 3NO_3^- + \frac{1}{2}I_2 + 6\ py$$

In $3[I_{py2}]\ NO_3$ and other similar compounds, I+ appears to be two-coordinate and hence $[Ipy^2]^+NO_3^-$ must be formulated as $[pyIpy]^+\ NO_3^-$.

Evidences for I_3^+

(i) Molten ICI_3 conducts electricity liberating iodine and chlorine at both electrodes. Thus, its ionisation is probably.

$$2ICI_3 \rightleftarrows ICI_2^+ + ICI_4^-$$

(ii) If iodine is oxidised by fuming nitric acid in acetic anhydride, I $(\text{acetate})_3$ is obtained. If this is electrolysed using silver electrodes, one equivalent of Agl is liberated at the cathode for the passage of 3 faradays of electricity. This confirms the ionisation.

$$I(\text{acetate})_3 \rightleftarrows I^{3+} + 3(\text{acetate})^-$$

Iodine acetate_3 forms colourless crystals, stable in the cold but decomposes at 100°C.

(iii) The iodine oxide l_4O_9 produced on treatment of iodine with ozonised oxygen can be formulated as $I^{3+}\ (IO_3)_3^{3-}$ by analogy with the above mentioned compounds.

(iv) When iodine is dissolved in oleum, I^{3+} ions are formed. If the oleum is concentrated, I^- is formed.

(v) Aubke (1995) spectroscopically confirmed the existence of I^{3+} ion by the reaction of iodine with HSO_3F,

Preparation of Compounds Containing I_3^+

(i) *Iodine Phosphate IPO_4 :* If iodine is oxidised by nirtic acid in the presence of acetic anhydride and phosphoric acid, iodine phosphate is obtained.

$$I_2 + \text{acetic anhydride} + \text{phosphoric acid} \rightarrow IPO_4$$

Iodine phosphate is a yellow crystalline solid, readily hydrolysed by water.

$$5IPO_4 + 9H_2O \rightarrow I_2 + 3HIO_3 + 5H_3PO_4$$

On electrolysis of aqueous solution of IPO_4, I_2 is liberated at cathode.

$$IPO_4 \rightarrow I^{3+} + PO_4^{3-}$$

$$2I^{3+} + 6e^- \rightarrow I_2 \text{ (At Cathode).}$$

(ii) *Iodine Perchlorate $I(ClO_4)_3$:* iodine perchlorate $I(ClO_4)_3 . 2H_2O$ is obtained in the form of yellowish green needles by the action of ozone on a cooled solution of iodine in anhydrous perchloric acid.

$$I_2 + O_2 + 6HClO_4 \rightarrow 2I(CO_4)_3 + 5H_3PO_4$$

The salt separates as $I(ClO_4)_3.2H_2O$. It is easily decomposed by water.

If silver perchlorate is added to excess iodine in ether at –85°C, a solution of yellow $I(ClO_4)_3$ is formed.

$$I_2 + 2AgClO4 \rightarrow AgI + AgCl_4.IClO_4$$

$$AgClO_4.\ IClO_4 + IClO_4 \rightarrow Agl + I(ClO_4)_3$$

(iii) *Iodine triacetate, $I(OCOCH_3)_3$:* Fuming nitric acid in the presence of acetic anhydride oxidises iodine producing the compound $I(OCOCH_3)_3$.

$I(OCOCH_3)_3$ reacts with methyl sulphuric acid to give $(CH_3SO_3)_3I$.

When a saturated solution of $I(OCOCH_3)_3$ in acetic anhydride is electrolysed, the quantity of silver iodide formed at a silvered platinum gauze cathode is in good agreement with faraday's calculation, assuming the presence of I^{3+}

$$I^{3+} + Ag + 3e^{-} \rightarrow AgI$$

(iv) *Iodobenzene dichloride C_6H_4ICl* : It is obtained as yellow crystals when dry chlorine is passed into iodobenzene dissolved in chloroform.

Very diluted hydroxide solution slowly decomposes it to iodosobenzene C_6H_4ICl.